The Green Economy in the Global South

The idea and practice of the 'green economy' is gaining momentum, coinciding with financial instability, and continued economic woe in the Global North, but generally more positive economic circumstances in the Global South. 'Green economic initiatives' in the Global South are multiplying, and include carbon payments, ecotourism, community-based wildlife management, sustainability certification initiatives, and offsets by mining companies exploiting new resources. These initiatives are reallocating resources, redefining inequalities, and redistributing the fortune and misfortune of participants of the green economy and those excluded from it. They have also led to resistance – locally, nationally, and transnationally – and to demands for alternatives to market-driven instruments and solutions, which are generally gaining strength and coherence. The articles included in this volume bring together a multi-disciplinary team of scholars from North and South to provide nuanced analyses of green economy experiences in the Global South – analysing the opportunities they provide, but also the redistributions they entail and the kinds of resistances they face. The ultimate aim of the collection is to provide a critical, but balanced, overview of the emerging green economy in the Global South and point the way to possible adjustments, alternatives or radical resistance, depending on different situations. This book was originally published as a special issue of *Third World Quarterly*.

Stefano Ponte is a Professor in the Department of Business and Politics, Copenhagen Business School. His research focuses on global value chains, the political economy of sustainability initiatives, and transnational environmental governance.

Daniel Brockington is Director of the Sheffield Institute for International Development at the University of Sheffield where he holds a research chair. His research covers issues of conservation and environmental policy, celebrity advocacy, and livelihood change.

ThirdWorlds

Edited by Shahid Qadir, *University of London, UK*

ThirdWorlds will focus on the political economy, development and cultures of those parts of the world that have experienced the most political, social, and economic upheaval, and which have faced the greatest challenges of the postcolonial world under globalisation: poverty, displacement and diaspora, environmental degradation, human and civil rights abuses, war, hunger, and disease.

ThirdWorlds serves as a signifier of oppositional emerging economies and cultures ranging from Africa, Asia, Latin America, Middle East, and even those 'Souths' within a larger perceived North, such as the U.S. South and Mediterranean Europe. The study of these otherwise disparate and discontinuous areas, known collectively as the Global South, demonstrates that as globalisation pervades the planet, the south, as a synonym for subalterity, also transcends geographical and ideological frontier.

For a complete list of titles in this series, please visit https://www.routledge.com/series/TWQ

Recent titles in the series include:

The Green Economy in the Global South

Edited by
Stefano Ponte and Daniel Brockington

LONDON AND NEW YORK

First published 2017 by Routledge

2 Park Square, Milton Park, Abingdon, Oxfordshire OX14 4RN
52 Vanderbilt Avenue, New York, NY 10017

Routledge is an imprint of the Taylor & Francis Group, an informa business

First issued in paperback 2018

Introduction, Chapters 1-3, Chapters 5-8 © 2017 Southseries Inc.
Chapter 4 © Michela Marcatelli

British Library Cataloguing in Publication Data
A catalogue record for this book is available from the British Library

ISBN 13: 978-1-138-29120-1 (hbk)
ISBN 13: 978-0-367-13351-1 (pbk)

Typeset in Times New Roman
by RefineCatch Limited, Bungay, Suffolk

Publisher's Note
The publisher accepts responsibility for any inconsistencies that may have
arisen during the conversion of this book from journal articles to book chapters,
namely the possible inclusion of journal terminology.

Disclaimer
Every effort has been made to contact copyright holders for their permission to
reprint material in this book. The publishers would be grateful to hear from any
copyright holder who is not here acknowledged and will undertake to rectify
any errors or omissions in future editions of this book.

Contents

Citation Information

The chapters in this book were originally published in *Third World Quarterly*, volume 36, issue 12 (December 2015). When citing this material, please use the original page numbering for each article, as follows:

Chapter 1
The Green Economy in the global South: experiences, redistributions and resistance
Dan Brockington and Stefano Ponte
Third World Quarterly, volume 36, issue 12 (December 2015) pp. 2197–2206

Chapter 2
Four discourses of the green economy in the global South
Carl Death
Third World Quarterly, volume 36, issue 12 (December 2015) pp. 2207–2224

Chapter 3
Tourism and the green economy: inspiring or averting change?
Melanie Stroebel
Third World Quarterly, volume 36, issue 12 (December 2015) pp. 2225–2243

Chapter 4
Suspended redistribution: 'green economy' and water inequality in the Waterberg, South Africa
Michela Marcatelli
Third World Quarterly, volume 36, issue 12 (December 2015) pp. 2244–2258

Chapter 5
Extractive philanthropy: securing labour and land claim settlements in private nature reserves
Maano Ramutsindela
Third World Quarterly, volume 36, issue 12 (December 2015) pp. 2259–2272

Chapter 6
Responding to the green economy: how REDD+ and the One Map Initiative are transforming forest governance in Indonesia
Rini Astuti and Andrew McGregor
Third World Quarterly, volume 36, issue 12 (December 2015) pp. 2273–2293

Chapter 7

The neoliberalisation of forestry governance, market environmentalism and re-territorialisation in Uganda
Adrian Nel
Third World Quarterly, volume 36, issue 12 (December 2015) pp. 2294–2315

Chapter 8

Inverting the moral economy: the case of land acquisitions for forest plantations in Tanzania
M. F. Olwig, C. Noe, R. Kangalawe and E. Luoga
Third World Quarterly, volume 36, issue 12 (December 2015) pp. 2316–2336

Chapter 9

Performativity in the Green Economy: how far does climate finance create a fictive economy?
Sarah Bracking
Third World Quarterly, volume 36, issue 12 (December 2015) pp. 2337–2357

For any permission-related enquiries please visit:
http://www.tandfonline.com/page/help/permissions

Notes on Contributors

Rini Astuti is a PhD student at the Victoria University of Wellington, New Zealand. Her research interests focus on the REDD+ (Reducing Emissions from Deforestation and Forest Degradation Plus) program.

Sarah Bracking is a Professor at the University of KwaZulu-Natal, South Africa. She holds the SARCHi Chair in Applied Poverty Reduction Assessment. Her current research explores the political economy of poverty, development, and climate finance and structures of power in Africa.

Daniel Brockington is Director of the Sheffield Institute for International Development at the University of Sheffield, UK, where he holds a research chair. His research covers issues of conservation and environmental policy, celebrity advocacy and livelihood change.

Carl Death is Senior Lecturer in International Political Economy at the University of Manchester, UK. He is a co-editor of the journal *African Affairs* (from 2015–), the top ranked journal in African Studies.

Richard Kangalawe is an Associate Professor at the University of Dar es Salaam, Tanzania.

Emmanuel Joachim Luoga is based at the Mbeya University of Science and Technology, Tanzania.

Michela Marcatelli is a PhD student at the International Institute of Social Studies of Erasmus University Rotterdam, the Netherlands.

Andrew McGregor is an Associate Professor in the Department of Geography and Planning, Macquarie University, Australia. His research falls under the broad umbrella of human geography, political ecology and development studies.

Adrian Nel is Senior Lecturer in Geography at the University of KwaZulu-Natal, South Africa. His interests lie in research and teaching about contemporary human-environment/landscape relations in Southern and Eastern Africa.

Christine Noe is Senior Lecturer at the Department of Geography, University of Dar es Salaam, Tanzania.

Mette Fog Olwig is Associate Professor at the Department of Social Sciences and Business, Roskilde University, Denmark.

Stefano Ponte is a Professor in the Department of Business and Politics, Copenhagen Business School, Denmark. His research focuses on global value chains, the political economy of sustainability initiatives and transnational environmental governance.

Maano Ramutsindela is a Professor in Environmental and Geographical Science, University of Cape Town, South Africa. His research interests are borders, regions, land reform and transfrontier conservation.

Melanie Stroebel is a post-graduate student at the University of Manchester, UK. Her doctoral research investigates climate change mitigation approaches of transnational tour operators.

The Green Economy in the global South: experiences, redistributions and resistance

Dan Brockington[a] and Stefano Ponte[b]

[a]*Institute for Development Policy and Management, University of Manchester;* [b]*Department of Business and Politics, Copenhagen Business School, Denmark*

As multiple visions for a Green Economy seek to become real, so are green economic initiatives in the global South multiplying. These can offer integration into wealth-generating markets – as well as displacement, alienation, conflict and opportunities for 'green washing'. The articles included in this collection bring together a multidisciplinary team of scholars and a range of case studies, from forestry governance to tourism to carbon finance, to provide nuanced analyses of Green Economy experiences in the global South – examining the opportunities they provide, the redistributions they entail and the kinds of resistance they face.

Introduction

The momentum gathering behind the idea and practice of the 'Green Economy' is coinciding with financial instability and continued economic woe in the global North, but generally more positive economic circumstances in the global South. Green Economy initiatives in the global South are multiplying, and include carbon payments, ecotourism, community-based wildlife management, sustainability certification initiatives and offsets by mining companies exploiting new resources. These are all part of a landscape offering new commodities, opportunities for commercialisation and possible integration into wealth-generating markets. But so too are growing incidents of land and water grabbing, displacement and alienation of resources required for wealthy tourists, conflicts over locally defined rules of access to carbon purchased by wealthy foreigners and instances of 'green washing' and other harmful activities. The Green Economy is reallocating resources, reinforcing inequalities and redistributing the fortune and misfortune of its participants and of those excluded from it. The articles included in this special issue bring together a multidisciplinary team of scholars to provide nuanced analyses of Green Economy experiences in the global South – examining the opportunities they provide, the redistributions they entail and the kinds of resistance they may face.

Behind the Green Economy lies a bundle of paradoxes and contradictions. The term is both a rallying call for radical change to the organisation of economic activity and social life, and an instrument by which meaningful alterations of either is resisted. It is a nascent idea, a future scenario, and yet projects in its name are transforming rural and urban realities in many locations in the global South. The Green Economy includes different epistemic communities, who seem to barely know each other, and is a field of vigorously contested interpretations. It demands attention, yet its breadth and multiple meanings make rigorous analysis hard. It is a fecund, sometimes febrile, arena of new scholarship.[1]

The growing vigour of the topic is plain. Brown and colleagues trace the rise in writing on the Green Economy from 12 papers in 2004–05 to over 50 times that number eight years later.[2] These works examine specific sectors in particular countries, advances in new technologies, discursive formations, political battles within international institutions, and blueprints for how to transition to green (or at least greener) economic activity. The addition to the breadth that this collection provides is its focus on the dilemmas the Green Economy throws up in diverse situations in selected economies, societies and political institutions in the global South.

We are not presuming to generalise across the Global South. Indeed, Carl Death's article specifically seeks to disaggregate the variety of responses to the Green Economy that can be found there. But we do see in these economies a series of different perspectives and patterns of practice that are more rarely encountered in the global North. For example, in Latin America more radical versions of the Green Economy are being explored, if only discursively, that entail looking beyond measures of mere GDP to determine prosperity. Elsewhere, and particularly in many African countries, moves towards a Green Economy are more modest and viewed with some distrust, precisely because they are perceived to threaten growth opportunities. As Faccer and colleagues make clear, economies that are built on extracting natural resources, are developing new coal deposits and producing more energy cheaply, and have large sections of their population seeking employment and ways out of poverty, will be less prone to consider potentially more expensive development trajectories.[3] Yet, at the same time, moves to promote particular forms of the Green Economy – ecotourism, carbon offsets and community-based natural resource management – find all sorts of new venues in which to emerge in various countries and communities of the global South. As the articles in this collection will show, despite governments' scepticism, moves towards a Green Economy have teeth and presence in these locations that are more visible than in the global North.

One of the reasons for the Green Economy's visibility in the global South is that tropical deforestation is seen as one of the low-hanging fruit that can be targeted rapidly to reduce carbon emissions. Such deforestation is one of the major causes of carbon emissions (it was at the time of the Stern report thought to be the most important); dealing with it was thought to be relatively cheap and certainly politically much easier than tackling, for instance, car use in industrialised countries. This would explain the remarkably consistent line that appears in a series of different celebrities' messages on behalf of conservation organisations, with all proclaiming that tropical deforestation causes more carbon emissions than 'all the cars, trucks and planes in the world combined'.[4] Thus landscapes and economies in the global South become vital actors in the defence of lifestyles in the global North. Another reason for the visibility and tangibility of the Green Economy in the global South is that practices such as payments for ecosystem services, green finance and ecotourism are materially manifest there. These practices result in new commodities, revenue streams, value chains, financial instruments and governance arrangements that are reshaping resource use and management at the local, and occasionally national and transnational, levels.

In addition to the actual Green Economy activities forming in the global South, there is also an important history of academic interest in their predecessor activities in this realm. Many of the measures that are now badged as Green Economy have a previous gestation in the form of a series of interactions between environmental affairs and the functioning of capitalism. These have been tackled by scholars writing about the 'neoliberalisation of nature' and neoliberal conservation, and contain many aspects (payments for ecosystem services, the use of markets to govern natural resources and environmental problems) that are now promoted as part of a Green Economy – this time in the context of growing acceptance of the role of the state in governing economic and environmental affairs. And yet, as is so often the case, authors and locations of this debate have been based mostly in the global North.

This had been notably the case among the epistemic community that lies behind the present collection, which gathered in 2008 at meetings in Manchester and Washington, DC, with subsequent gatherings in Lund, Sweden, The Hague and Toronto. These were all lively, vibrant conferences and workshops with vigorous debate and important publications deriving from them.[5] They became larger with each iteration, and helped to successfully develop critical academic thinking in this area – together with other efforts at the meetings of professional associations of geographers, anthropologists and development studies in the USA and UK. But these were not South-based endeavours.

Our point in observing this imbalance is not that intellectual endeavours about the South have to be written from it, or by its residents.[6] We do not subscribe to such simplistic notions of authenticity or authority. The freedoms of international academia mean that we are not bound by the contingencies of location in our research. But international intellectual agendas are likely to work better if they involve different parts of the world more equally. If the limitations of funding and visas constrain intellectual networks, then the communities resulting are stunted. It matters particularly for any scholars who are trying not just to analyse for intellectual audiences, but also to engage and work with the communities about whom they write. Those sorts of interactions are harder if so much of the intellectual activity is in a different hemisphere, and if the academic communities in which they take part are similarly biased.

The articles included in this collection derive from an attempt at least partially to redress this imbalance. Earlier versions of these papers were presented at the 'Green Economy in the South' conference, which was held in July 2014 at the University of Dodoma in Tanzania, and hosted by its Department of

Geography and Environmental Sciences. Once again, our point is not that this makes the meeting any more authentic or authoritative. Indeed, it became plain that there are far more enduring inequalities at work than merely hosting a meeting in Dodoma can overcome. It proved too hard to arrange travel and visa documents for applicants based in the Democratic Republic of Congo. Applications from North-based scholars were still more numerous than those based in the South and, even if it is closer to many potential participants, Dodoma can still be hard and expensive to get to. To the extent that our goal of hosting a more inclusive conference was achieved, it was only possible because of generous sponsorship. Nevertheless, the result was that over 70 people attended from nearly 30 countries. Because of the high quality of the venue, the strong media coverage the conference attracted, and the measures the hosts took to welcome and cater for delegates' needs, this meeting did succeed in providing a vigorous and lively international conference with broader geographic participation.

The articles included in this collection are but a fraction of those presented at the conference. They provide a number of critical perspectives – focusing on what kinds of power realignments and redistribution of resources take place under the aegis of the Green Economy, and examining the extent to which support and resistance play out in a variety of settings and who is involved. They also help to demonstrate some of the existing research gaps and suggest future directions required by the epistemic community that has produced them, which we briefly address at the end of this introduction.

The articles

Carl Death's article, 'Four discourses of the Green Economy in the global South', sets the scene by disaggregating the different discourses of the Green Economy predominant in selected countries in the global South. Death argues that, while the concept is often deployed in a consensual and win-win manner, we need to understand Green Economy discourses as part of a broader political economy of the 'green state' in the global South. The first of four discourses he highlights is 'green resilience', which seeks ways to strengthen local and national economies to cope with the threats of climate change while delivering economic growth (with Ethiopia as the main example). The second, visible in China and India, is 'green growth', in which the transformations required to make economies greener are embraced as potential means of increasing the volume of transactions and economic activity. The third, 'green transformation', exemplified by aspects of South Korea's green stimulus policies, refers to states seeking to invest considerable resources into reshaping their economies to make them greener.[7] The fourth and last, 'green revolution', invokes the possibility of more radical change, and possibly even de-growth, which the author has observed especially in Brazil. Tracing this variety allows Death to highlight important commonalities: first, that the state is important in shaping and directing discourse and policies – even social movements' radical discourses often seek recognition and support from state institutions; and, second, that all these discourses are compatible with perpetuated inequalities.

Three articles included in this collection critically examine the role of tourism in the Green Economy. Melanie Ströbel's article, 'Tourism and the Green

4

Economy: inspiring or averting change?', examines the efforts of the tourism industry to promote green growth. Tourism matters a great deal because it is one of the main ways in which ecosystem services can be paid for – wealthy tourists are prepared to part with significant sums to protect landscapes, water supplies, forests and so on. The industry can provide an effective means of directly transferring resources from richer visitors to poorer residents. Ströbel shows clearly that advocates of the tourism industry do not hesitate to claim those benefits. They also claim that tourism needs to increase in order to provide more and better benefits. And yet these increases, Ströbel argues, normally entail substantial growth in CO_2 emissions. There is optimism in the official documents that she analyses, suggesting that low carbon growth is possible, but little reason to put faith in that optimism. In a dispassionate, but depressing analysis, Ströbel shows how supporters of this aspect of the Green Economy seem to have unrealistic expectations.[8] Tourism is promoted as a strategy for solving problems, but the problems with which it is intricately bound are not mentioned, and more radical possibilities – such as shrinking tourism – simply do not feature. This limits the political space for promoting more demanding possibilities of change.

The other two authors examine different tourism and conservation initiatives in South Africa. Michela Marcatelli, in her article entitled 'Suspended redistribution: Green Economy and water inequality in the Waterberg, South Africa', shows that conservation initiatives in South Africa interact with local needs not just through providing jobs and tourist revenues, and by claiming land, but also through the water resources that they command materially and discursively. This takes place in the context of a country that has seen insufficient moves to redistribute land and the continued removal of rural populations from once 'white farms' to new (often also white-owned) game reserves.[9] People, and people's needs, have moved from rural areas to nearby urban centres. But the water (and especially its infrastructure and accessibility) has not flowed with them. Instead, it remains privately accessed, and therefore restricted, behind the fences of commercial and private game farms. 'Green' game farms – land uses that promote both natural environments and economies built on sustainable use of natural resources – become instrumental in perpetuating unequal social divisions of resources.

Maano Ramutsindela makes a similar argument. His article, 'Extractive philanthropy: strategies for securing labour for private nature reserves', examines philanthropic management and promotion of tourist businesses on private game farms in South Africa. He argues that post-apartheid South Africa provides conditions under which a seamless connection between philanthropy, labour and land claims in private nature reserves emerges. Philanthropy allows private owners to structure and control labour while directly or indirectly affecting the trajectory of land claims in the area. As a result of this, private game farms have been able to oppose and restrict land restitution in the name of promoting businesses that will benefit neighbours and formerly dispossessed communities. The result is portrayed, no doubt, as a win-win. Some jobs are created and much income is produced from these high-end eco-tourism resorts. But his critique is as much about the road not travelled by these choices. For the major winners are the wealthiest elites, who can afford to manage these reserves or frolic within them. Once again the sustainable Green Economy built on (this variety of) eco-tourism is riven with inequality.

The next three papers examine how Green Economy principles, policies and discourses are affecting forest policies. In 'Responding to the Green Economy: how REDD+ and the One Map Initiative are transforming forest governance in Indonesia', Rini Astuti and Andrew McGregor examine the ways in which Indonesian forests are becoming more legible for the purpose of mapping and securing carbon offsets. The issue here is that official maps and understandings of who owns what forest and who has concessions to do what are often inconsistent and inaccurate, allowing all sorts of problematic forest development to occur. The One Map Initiative was the government's response to this messiness by providing a single accurate record. Astuti and McGregor argue this is a technical fix, and as such not necessarily oppressive. It can provide a means of disciplining the powerful. But, as with all official records, one has to ask: what can the state see? And what must local forest users do (to themselves) in order to be seen by the state? Astuti and McGregor show that there are forms of forest, types of forest user and processes of registration of interests, presence and rights in this initiative that are far from equitable. Local groups and NGOs are contesting these measures and making use of them for their own agendas, including attempts to address historical inequalities. The measures necessary to introduce Green Economy commodities (carbon offsets) require rendering forest societies and landscapes legible in ways that could profoundly alter life there, presenting a series of opportunities bundled up with specific risks.

Adrian Nel's article, 'The neoliberalisation of forestry governance, market environmentalism and re-territorialisation in Uganda', provides a detailed account of the reorganisation and restructuring of forestry in Uganda to allow space for market environmentalism. The result has been that forestry has become increasingly concerned with expanding plantations, and with criminalised local clearance of trees and forests. Additionally, Nel shows that the reorganisation of forestry is embedded in deeper reorganisations and contests within the Ugandan state itself. This has resulted in different forms of 'governance-beyond-the-state' that control, limit and reshape both behaviour and forest cover. As in Rini and McGregor's account in Indonesia, the advent of these forms of market governance in Uganda is accompanied by profound reorganisations of state governance, the social organisation of forestry, as well as social interactions with forests.

Mette Olwig, Christine Noe, Richard Kangalawe and Emmanuel Luoga examine in detail the case behind the establishment of plantation forests in Mufindi District, Tanzania. Their article, 'Inverting the moral economy: the case of land acquisitions for forest plantations in Tanzania', shows that these forests were planted by investors over thousands of hectares in two villages and represent significant forms of land use change and, potentially, restriction of access to residents. The authors argue that discursively the investors seize the moral high ground, because trees are 'green' (therefore good), the land underneath them is seen as idling and thus not productive, and their investments bring a valued boost to the local economy. This is possible even if the actual links and benefits to existing local economies are not clear, or may even be detrimental to local farmers. These plantation forests create an environment where change is presumed to be good, and where the evidence for that belief becomes less important. Creating such forests also means that villagers lose 30% of their land, moving closer to a cash economy, with the moral economy that had sustained them subverted to the demands of a 'global moral economy' that needs more trees. Moreover, in a remarkable twist to the story, villagers themselves are planting more trees, sometimes to the apparent detriment of their food security (as they lose cropland) because of the purported benefits trees can bring.

Sarah Bracking's contribution, 'Performativity in the Green Economy: how far does climate finance create a fictive economy?' steps back from the detail of particular projects and reorganisations of resources and considers the means by which supposedly Green Economy projects are financed, and what sort of change this will produce in the structure and performance of economies in the Global South and the global economy at large. In a careful, but chilling, account she shows how the evaluative and calculative devices used to invest in the Green Economy, and to mitigate and offset current pollution, can be remarkably divorced from realities on the ground. Indeed, they may even depend on being so. Value is created through diverse performances and appearances of being 'green', and not because of material change. This is not mere superficial behaviour attributable to individual companies, or 'green washing'. It is a much more far-reaching 'green cleansing' that pervades an entire sector.

These are detailed and careful articles examining diverse experiences and trajectories of the Green Economy in the global South and the redistributions they entail: (1) between different groups of actors accessing specific resources – mostly to the advantage of existing elites and to the detriment of weaker groups; (2) between different kinds of benefits, eg when the creation of 'Green Economy jobs' justifies land access limitations for local communities; (3) between private and public actors, eg when water accessibility for private game farms comes at the cost of marginalised urban communities; (4) between local and global outcomes, eg when income and conservation benefits from ecotourism are accompanied by global increases in CO_2 emissions; and (5) between different kinds of benefits, eg when environmental and economic benefits accrue to different groups of actors.

These redistributions are justified, normalised and 'naturalised' by more or less subtle transformations in moral and normative orders, eg when the establishment of forest plantations detrimental to local subsistence supports a global (environmental) moral economy. As Olwig et al in this collection put it, 'instead of having the upper moral ground, as poor people who have the right to a livelihood that can support their families, the poor are burdened with the moral responsibility of compensating for the excessive consumption of the more well-to-do in the global South and particularly in the global North'.

These articles show that Green Economy initiatives are often based on technologies and forms of spatial knowledge that tend to portray changes in access to resources in win-win and non-confrontational fashion. They bring in investors in the name of providing local benefits, but also establish new venues for philanthropy. They entail specific forms of financing, derivatives trading and financial speculation that are based on valuation performed within a virtual framing of 'care'. These initiatives may be accompanied by profound changes in political institutions and social relations, sometimes resulting in the strengthening of the bureaucratic and coercive capacity of the state (often in coalition with local elites and outside investors) at the expense of disenfranchised communities. And they lead to new patterns of financial, asset and resource accumulation based on green credentials and justifications – paradoxically including accumulation by offshore fossil-fuel and infrastructure funds based on the provision of international public subsidies (eg through CDM projects).

While some instances of resistance are reported in the articles included in this collection, these are often thwarted on moral, technological and economic grounds, and on the basis of the primacy of expert solutions and valuations. In only one case in Indonesia a Green Economy initiative was used as an opportunity to attempt to address historical inequalities in access to resources. In the name of continued growth, the Green Economy becomes a discursive strategy and a moral order backing up concrete practices that often reproduce existing inequalities in global and local political economies – marginalising the possibility for more radical transformations, such as those based on modifying lifestyles and *decreasing* consumption and travel.

These articles were built on many months of intensive doctoral research and analysis, or even longer periods of repeated encounters by more seasoned academics. They are typical of the standards and quality of work presented at the conference. The presenters who came to the conference were highly interested in natural resources, wildlife, water, trees, REDD and conservation. This focus is surprising. It means that at a conference that was meant to be about the Green Economy in the global South there were no papers on manufacturing or service industries. The call for this conference had been circulated globally. We received over 100 abstracts and invited over 70 speakers. Yet somehow the epistemic communities to which it appealed did not include people working on these relevant issues. For a critical community intent on critically examining capitalism, the economic sectors it chose to examine were limited.

Shoreman-Ouimet and Kopnina have recently argued that 'most environmental social scientists, for instance, do not study those groups that do the greatest damage (eg intensive agriculture, logging companies, chemical manufacturers, etc), where they could arguably make the greatest difference'.[10] Although this critique is off the mark, as there is a great deal of environmental scholarship which tackles these aspects (see, for example, the articles reviewed in Castree[11]), the epistemic groups working on the Green Economy in Dodoma were indeed narrow. Our scholarship needs to form better links with critics working in urban and industrial fields. Our point here is not to criticise the work of individual scholars whose focus is on rural areas. The majority of many countries in the global South remain rural and will do so for years to come, and good scholarship can require that sort of focus on specific sectors. Rather, our closing point is that there need to be better interactions and network building with the broader epistemic community. By collaborating more effectively with scholars working on 'the urban' or with particular industrial sectors, we will be able to analyse the Green Economy more effectively.

Acknowledgements

Previous versions of the papers included in this collection were presented at the 'Green Economy in the South' conference, which was hosted by the Department of Geography and Environmental Studies, University of

Dodoma, Tanzania, 8–10 July 2014. The conference was co-hosted by the Institute for Poverty, Land and Agrarian Studies (PLAAS), University of the Western Cape; the Institute for Development Policy and Management (IDPM), University of Manchester; the International Institute for Social Studies (ISS), Erasmus University; the Sustainability Platform, Copenhagen Business School (CBS); and the Future Agricultures Consortium. In particular, we would like to thank the other members of the organising committee for their invaluable work: Davis Mwamfupe, Abiud Kaswamila, Thabit Jacob, Mathew Bukhi and Wilhelm Kiwango (University of Dodoma); Emmanuel Sulle (PLAAS); Sarah Bracking (IDPM); Bram Büscher (University of Wageningen); Jim Igoe (University of Virginia); Baruani Mshale (CIFOR); and Christine Noe (University of Dar es Salaam).

Funding

Funding that made the 'Green Economy in the South' conference possible was kindly provided by UNEP, the African Studies Association (UK) and the Sustainability Platform, Copenhagen Business School.

Notes

1. Bailey and Caprotti, "The Green Economy"; Bauhardt, "Solutions to the Crisis?"; and Tienhaara, "Varieties of Green Capitalism."
2. Brown et al., "Green Growth or Ecological Commodification."
3. Faccer et al., "Interpreting the Green Economy."
4. For example, the actor Harrison Ford, https://www.youtube.com/watch?v=uVBw66t8a9Q; Prince Charles (UK), https://www.youtube.com/watch?v=boEDMVNAPk4; and journalist Tom Friedman, https://www.youtube.com/watch?v=W1Li3O81uDs, all accessed July 29, 2015.
5. Brockington and Duffy, *Capitalism and Conservation*; Arsel and Büscher, "Nature[TM] Inc"; and Büscher et al., *Nature[TM] Inc*.
6. This imbalance has been raised before, though not fully settled. See, for example, Sidaway, "In other Worlds."
7. But Death is careful to report the objections that Tienhaara raises to using such investments to dam and dredge rivers.
8. Wanner, "The New 'Passive Revolution' of the Green Economy."
9. Connor, "Opportunity and Constraint."
10. Shoreman-Ouimet and Kopnina, "Reconciling Ecological and Social Justice," 324.
11. Castree, "Neoliberalizing Nature: Processes, Effects and Evaluations"; and Castree, "Neoliberalizing Nature: The Logics of De- and Re-regulation."

Bibliography

Arsel, M., and B. Büscher. "Nature™ Inc: Changes and Continuities in Neoliberal Conservation and Market-based Environmental Policy." *Development and Change* 43, no. 1 (2012): 53–78.

Bailey, I., and F. Caprotti. "The Green Economy: Functional Domains and Theoretical Directions of Enquiry." *Environment and Planning A* 46 (2014): 1797–1813.

Bauhardt, C. "Solutions to the Crisis? The Green New Deal, Degrowth, and the Solidarity Economy – Alternatives to the Capitalist Growth Economy from an Ecofeminist Economics Perspective." *Ecological Economics* 102 (2014): 60–68.

Brockington, D., and R. Duffy. *Capitalism and Conservation*. London: Wiley, 2011.

Brown, E., J. Cloke, D. Gent, P. H. Johnson, and C. Hill. "Green Growth or Ecological Commodification: Debating the Green Economy in the Global South." *Geografiska Annaler: Series B, Human Geography* 96, no. 3 (2014): 245–259.

Büscher, B., W. Dressler, and R. Fletcher. *NatureTM Inc: Environmental Conservation in the Neoliberal Age*. Tucson, AZ: University of Arizona Press, 2014.

Castree, N. "Neoliberalizing Nature: The Logics of De- and Re-regulation." *Environment and Planning A* 40 (2007): 131–152.

Castree, N. "Neoliberalizing Nature: Processes, Effects and Evaluations." *Environment and Planning A* 40 (2007): 153–171.

Connor, T. K. *"Opportunity and Constraint: Historicity, Hybridity and Notions of Cultural Identity in the Sundays River Valley (Eastern Cape) and Pafuri (Mozambique)"*. PhD Thesis., Rhodes University, 2006.

Faccer, K., A. Nahman, and M. Audouin. "Interpreting the Green Economy: Emerging Discourses and their Considerations for the Global South." *Development Southern Africa* 31, no. 5 (2014): 642–657.

Shoreman-Ouimet, E., and H. Kopnina. "Reconciling Ecological and Social Justice to Promote Biodiversity Conservation." *Biological Conservation* 184 (2015): 320–326.

Sidaway, J. D. "In Other Worlds: On the Politics of Research by 'First World' Geographers in the 'Third World'." *Area* 24, no. 4 (1992): 403–408.

Tienhaara, K. "Varieties of Green Capitalism: Economy and Environment in the Wake of the Global Financial Crisis." *Environmental Politics* 23, no. 2 (2014): 187–204.

Wanner, T. "The New 'Passive Revolution' of the Green Economy and Growth Discourse: Maintaining the 'Sustainable Development' of Neoliberal Capitalism." *New Political Economy* 20, no. 1 (2015): 21–41.

Four discourses of the green economy in the global South

Carl Death

Politics, University of Manchester, UK

This article identifies four contrasting global discourses of the green economy in contemporary usage: green resilience, green growth, green transformation and green revolution. These four discourses are manifested in recent green economy national strategies across the global South, including in Ethiopia, India, South Korea and Brazil. Disaggregating these discourses is politically important, and shows their different implications for broader political economies of the green state in the global South.

Introduction

The 'green economy' is the latest repackaging of long-running debates, programmes and discourses ostensibly seeking to reconcile economic growth and capitalist development with ecological sustainability. As such, much like the discourse of sustainable development before it, advocates of a green economy tend to present it as an unquestionable good: who could possibly oppose an economy which is 'low carbon, resource efficient, and socially inclusive'?[1] In contrast, critics have labelled the green economy discourse contradictory, distracting or politically dangerous, legitimating new forms of expropriation and accumulation.[2] The terrain of debate is thus very similarly to earlier debates over sustainable development, which for some was the best hope of combining environmental, social and economic values, whereas others saw the discourse as 'polite meaningless words' masking continued capitalist exploitation.[3] There seems to be an intractable opposition here, a fundamental disagreement over the politics of the green economy. In some senses this is correct, particularly in terms of attitudes towards the sustainability of capitalism as a system. However, the very stark disagreements over the politics of the green economy are also a product of the competing and contradictory discourses which that term conceals, and which this article seeks to illuminate. The green economy means quite different things to different people and in different contexts and the article maps four discourses of the green economy, derived from key articulations by the United Nations Environment Programme (UNEP), the United Nations Development Programme (UNDP), the World Bank, the Green Economy Coalition and others. These are the discourses of green resilience, green growth, green transformation and green revolution.

The article then explores how these discourses are manifested in actual examples of national green economy strategies in the global South. Doing this provides a powerful illustration of the political significance of the type of green economy discourse adopted by governments, as these discourses have quite different political implications. Moreover, this also broadens analysis of the green economy beyond cases of 'ecological modernisation' in developed, industrial, countries in the global North. Germany, Japan and Scandinavia tend to be the most common examples of countries which are turning towards high-tech industries and renewable energy sources, and implementing the latest 'green' designs in transport and construction.[4] It is in these countries where some theorists have seen the greatest potential for the emergence of 'green states', the latest stage of modernisation after the liberal democratic state and the welfare state.[5] Yet it is plausible to imagine that green economy strategies are manifested in different forms in the global South, where discourses of sustainable development and the developmental state have been more prominent. The relative absence of in-depth analysis of green economy strategies in the global South means that our current understanding of the range of political interventions under the green economy banner is somewhat restricted.[6]

The second reason why a focus on national green economy strategies is important is that most of the existing literature on the green economy in the global South has tended to focus on rural or agricultural dimensions of the green economy: natural resource conservation schemes, forestry projects like REDD+, agricultural investments and biofuels, and payments for ecosystem services.[7] Moreover, informed by political ecology approaches, most of these studies have provided close analysis of particular local sites and cases.[8] This research is extremely important, and is essential in unpacking the local contradictions, complications and discourses at work. But there is an absence of serious consideration of the national strategies and developmental programmes being deployed by Third World states, some of which are mobilising the green economy in ways which have only peripheral relationships to the traditionally 'green' issue areas of conservation and natural resource management.[9]

This, then, is the most important contrast with the discourse of sustainable development which dominated environment–development debates 20 years ago: the green economy is a firmly statist concept, in which states are the primary architects and subjects of political interventions. This is true across the discourses of green resilience, green growth, green transformation and green revolution, as we will see below. Even radical and revolutionary discourses, driven by social movements, seek recognition, concessions and reforms from state institutions and political actors. This statism is a product of the coalescing of discourses of developmental states, elite postcolonial nationalist projects, and a broader re-legitimation of 'unprecedented public interventions into economic life' following the global financial and economic crises of 2007–08.[10] The emergence of new forms of green states in the global South is thus a key site of academic enquiry and political activism in struggles over competing discourses of the green economy.[11]

According to UNEP, a green economy is one that results in 'improved human well-being and social equity, while significantly reducing environmental risks and ecological scarcities'.[12] This article is not about how to define these terms more precisely, or determining which policies are most likely to achieve UNEP's vision. Rather, it is an assessment of the ways in which the green economy discourse has been mobilised in the global South, and the political consequences of some of these mobilisations. The next section considers what a discursive approach to the green economy entails, before subsequent sections map out the green economy through four discourses of green resilience, green growth, green transformation and green revolution. These discourses are then examined in the contexts of national strategies in various countries, with a focus on Ethiopia, India, South Korea and Brazil. The concluding argument is that what these discourses of the green economy have in common is a re-legitimation of the production of strong 'green states' in the global South: states which are seeking to govern markets, borders, people and international relations in new and sometimes authoritarian ways. As James Meadowcroft concludes, 'as the decades have advanced, the state has been forced to accept that an ever more profound transformation of economic activity and of political and legal obligations will be required if environmental problems are to be managed'.[13]

Discourses of green economies and green states

The concept of a 'green economy' was reinvigorated in public policy discussions after 2008, when UNEP launched its Green Economy Initiative and other international institutions, such as the OECD and the World Bank, enthusiastically promoted the idea. But, as World Bank economists have pointed out, proposals on how to make the economy greener can be found in economics 'textbooks going back at least to the 1950s...with environmental taxation, norms, and regulations being the main tools of a green growth strategy'.[14] Interest in environmental economics burgeoned in the 1980s, alongside the concept of sustainable development, and the work of researchers such as David Pearce and his co-authors attracted high-level attention, for example in their 1989 report for the UK government, *Blueprint for a Green Economy*.[15] The central idea was that a green economy would be one in which environmental externalities would be fully accounted for, resting on the assumption that free resources would inevitably be exploited, depleted or polluted. In 2012 two of Pearce's original co-authors, Edward Barbier and Anil Markandya, published *A New Blueprint for a Green Economy*, in which they argued that the message of 1989 is still relevant, and that the threefold challenges of valuing the environment, accounting for the environment, and incentives for environmental improvement, are more urgent than ever. As Barbier explains, 'we use our natural capital, including ecosystems, because it is valuable, but we are losing natural capital because it is free'.[16]

If the arguments of the environmental economists have changed little since the 1980s, the political and economic context of the early twenty-first century is very different. Two crises are crucial to understanding why the idea of the green economy has returned to prominence: the climate crisis and the financial crisis.[17]

The 2000s was the decade when climate change became a salient international issue: the Kyoto Protocol came into force, it was discussed at the UN Security Council and Nicholas Stern made the case for the economics of climate change in as clear and compelling a manner as is possible: action later will be more costly than action now.[18]

Enthusiasm for a green economy was further fuelled by the interlocking financial and economic crises which began in 2007 and 2008, and which rocked the US and European economies and then the rest of the world.[19] UNEP's key report on the *Global Green New Deal* begins by noting that 'the world today finds itself in the worst financial and economic crisis in generations'.[20] Its diagnosis of the interconnected financial, economic, social and environmental crises became even more explicit in a later publication.

> The causes of these crises vary, but at a fundamental level they all share a common feature: the gross misallocation of capital. During the last two decades, much capital was poured into property, fossil fuels and structured financial assets with embedded derivatives. However, relatively little in comparison was invested in renewable energy, energy efficiency, public transportation, sustainable agriculture, ecosystem and biodiversity protection, and land and water conservation.[21]

For the UNEP authors the solution is clear: greater public regulation of the economy. 'To reverse such misallocation requires better public policies, including pricing and regulatory measures, to change the perverse incentives that drive this capital misallocation and ignore social and environmental externalities.'[22] Together climatic and financial crisis have re-legitimated demands for a new, cleaner, greener economy.

The return of the notion of the green economy has been accompanied by a growing acceptance of the role of the state in governing economic transitions and transformations.[23] The post-Washington Consensus on the need for better political regulation of markets has held;[24] 'big development' is back in fashion, with development banks and state agencies funding large infrastructure investments;[25] and most environmentally minded commentators are calling for 'more rather than less state intervention in the economic processes of investment, production, and even consumption'.[26] States were required to bail out the banks in the global North, and the concept of the developmental state has been extended beyond East Asia to Africa and Latin America. Recent articulations of the developmental state literature have emphasised the importance of responding to ecological limits and environmental degradation.[27] These trends appear to contradict the assumption in ecological modernisation literatures on the 'green state' – defined as a state in which ecological imperatives have become central to state actions – that the most important progress towards environmentally orientated state action will come from the global North.[28] National strategies for the green economy thus build upon long-standing discourses of state institutions as tools for social and economic progress among postcolonial nationalist elites, who saw the state as 'the chisel in the hands of the new sculptors'.[29] As Michelle Williams suggests, 'countries such as Bolivia and Ecuador are pioneering possible approaches to nature that require us to rethink our understanding of development and demonstrate once again that ideational shifts are emerging from the Global South'.[30]

This article argues that these green economy discourses produce politically significant spheres of governable activity, forms of knowledge and governing techniques, and new actors and subjectivities. Taking a discursive approach means taking seriously the language, frames, world-views and assumptions of the statements which comprise the green economy discourses, rather than dismissing them as meaningless, simple 'green-wash' or 'polite, meaningless words'.[31] Discourses constitute certain ways of thinking about, representing and acting upon the world.[32] Within discourses particular things are made visible and others invisible, truths are created and regimes of knowledge established, practices and technologies are concretised and subjects are produced. Material objects may exist independently of discourse, but it is discourses which give them meaning and significance.[33] Discourses are systems of representation that produce meaning itself, or 'practices that systematically form the objects of which they speak'.[34]

Sustainable development is a good example of a discourse which, in broad terms, argues that it is possible to have economic growth, environmental sustainability and social responsibility in a win-win-win form of development – but this discourse is also bound up with the production of an array of subjects (such as responsible civil society partners), technologies (such as environmental impact assessments), sites (such as intergovernmental summits) and forms of knowledge (such as environmental economics).[35] While at one level it could plausibly be argued that environmentalism is a discourse, it is more accurate to claim that contemporary environmentalism involves a range of discourses. John Dryzek, for example, identifies discourses of limits, administrative and economic rationalism, democratic pragmatism, sustainable development, ecological modernisation, green consciousness and ecological democracy.[36] In the following section four discourses of the green economy are delineated by drawing attention to the different ways in which they produce politically significant spheres of governable activity, forms of knowledge and governing techniques, and new actors and subjectivities.[37]

Four discourses of the green economy

These four discourses are derived from key articulations of the green economy produced by UNEP, UNDP, the World Bank and the Green Economy Coalition.[38] Of course, other analysts have disaggregated the green economy in different ways. Kyla Tienhaara, for example, distinguishes between the 'Green New Deal', 'Green Stimulus' and 'Green Economy' variants.[39] Peter Ferguson identifies the weak green economy, transformational green economy and strong green economy.[40] These terms tend to be deployed by different actors in different ways at different times, however; hence it is more useful to try to clearly distinguish four discourses of the green economy that often overlap in practice, but which represent clearly identifiable competing priorities when disaggregated. Here I identify discursive trends from key articulations of the green economy, and then explore how these are manifested in specific national strategies. The terms 'green resilience', 'green growth', 'green transformation' and 'green revolution', are therefore those used by governments and movements themselves.[41]

Green resilience

The first discourse is essentially a defensive reaction to the crises outlined above: an attempt to ensure the sustainability and stability of the economy and social life in the face of peak oil, changing climates, food and water insecurity and loss of biodiversity. This discourse brings 'the climate crisis' and 'the environmental crisis' into stark visibility, together with their impact on vulnerable societies.[42] UNEP notes the importance of the 'readiness and resilience of an economy and population to cope with change' for achieving a green economy,[43] while the World Bank calls for a green economy that is 'resilient in that it accounts for natural hazards and the role of environmental management and natural capital in preventing physical disasters'.[44] A World Bank programme launched in 2009 entitled *Making Development Climate Resilient in Sub-Saharan Africa* sought to provide developmental assistance to African states, including 'making adaptation and climate risk management a core developmental component'.[45]

A resilient green economy is to be achieved by a combination of technocratic interventions by states and development institutions, together with empowered communities who (it is hoped) can draw on their own sources of resilience.[46] Thus dominant forms of knowledge include developmental disciplines like crop science, water and sanitation engineering, social psychology and price monitoring for food markets.[47] Some articulations of the discourse allow more or less prominence for forms of indigenous knowledge.[48] Key techniques include developmental projects at the regional or local scale, intended to 'prop up' the central subjects of green resilience, namely vulnerable communities. The UNDP, for example, supports projects including targeted cash transfers in Ethiopia, weather-based crop insurance in Malawi and asset transfers in Bangladesh.[49]

The discourse of resilience in the face of environmental risks and hazards can be identified in many national development strategies across the global South, including South Africa's 'National Climate Change Response', Rwanda's 'Green Growth and Climate Resilience' strategy, Dominica's 'Low-carbon Climate-resilient Development Strategy' and Bangladesh's 'Reducing Vulnerability to Climate Change Project'. One of the most high-profile examples is Ethiopia's 'Climate-resilient Green Economic Strategy', which was launched in 2011 and aims for zero net greenhouse gas emissions; reduced vulnerability to climate change-associated risks; and for the country to become a 'green economy frontrunner' by investing in low carbon infrastructure.[50] Ethiopia continues to be associated with hunger and drought in many Western minds and newspapers since the catastrophic famines of the 1970s and 1980s, and more recent crises in 1999–2000 and 2002–03.[51] Yet Ethiopia has also been one of the fastest growing economies of the 2000s, averaging 10.7% GDP growth per year. In 2012 Ethiopia was the 12th fastest growing economy in the world.[52] The 'green economy' discourse is thus one element of a broader strategy to reposition the country in both cultural and economic global political imaginaries. The central element of the Climate-resilient Green Economic Strategy is a massive increase in hydropower generation through dam construction, with a projected growth in electricity capacity from 7 TWh (terawatt hours) to 80 TWh by 2030, of which 90% will be provided from hydropower.[53] Yet there is also a number of programmes specifically designed to increase resilience, for example through micro-insurance.[54]

UNEP notes that 'enabling the poor to access micro-insurance coverage against natural disasters and catastrophes is important for protecting livelihood assets from external shocks due to changing and unpredictable weather patterns'.[55] For example, index insurance, which makes payments when rainfall levels fall below a certain level has been trialled among small-scale farmers in Ethiopia.[56] In combination with larger and more established programmes – such as the Productive Safety Net Programme (PSNP), a government-run, USAID-funded programme through which residents of the community work as labourers – vulnerable small-scale farmers are able to survive irregular weather patterns and other periods of dearth.[57] For reasons like this micro-insurance (along with other techniques of micro-credit and micro-finance) has been portrayed 'as the new green revolution'.[58]

However, even such potentially positive interventions have a number of unintended or undesirable consequences. As Petersen's study indicates, the provision of crop insurance to smallholder farmers can be viewed by the private sector as a way of accessing a previously untapped market: 'insurers involved in the Ethiopia index insurance anticipate offering health and automobile insurance to the participants and banks envision larger loans to increasingly wealthy farmers in Malawi', creating a 'culture of insurance' and further entrenching the monetisation and financialisation of rural agriculture.[59] Moreover, Ethiopia's overall climate resilience strategy is a largely top-down project driven by an autocratic president; it can be seen in a longer history of state-led, hierarchical and often coercive modernisation projects.[60] Green resilience discourses can never be confined to technocratic programmes and they always have political impacts. While the discourse of resilience is not necessarily conservative, and indeed was developed by political ecologists to build upon existing vulnerability approaches by emphasising the adaptive capacity and agency of subaltern populations towards progressive ends,[61] its manifestation as part of national green economy strategies often seems to involve strengthening the bureaucratic and coercive capacity of state institutions at the expense of local communities.

Green growth

The second discourse is green growth, which has been the dominant global form of the green economy since the financial crisis. In this discourse environmental changes and programmes are primarily viewed as an economic opportunity, not a threat. As the World Bank report *Inclusive Green Growth* argues, the 'current system is inefficient, thereby offering opportunities for cleaner (and not necessarily slower) growth'.[62] Rising world populations, and rapidly growing economies in Asia, Latin America and Africa, also seem to present new markets. Even in older markets the niches for 'organic' and 'green' products are becoming increasingly commercially attractive.[63] As Büscher and Fletcher put it, 'Green is hot', and various techniques and technologies such as carbon markets, bio-prospecting, and commodifying ecosystem services come together in what they term 'accumulation by conservation': 'potentially a new "phase of capitalism" as a whole, imbued with a productive form of power that shapes new joint environmental and accumulation possibilities'.[64] The dominant form of knowledge in this discourse is mainstream neoclassical economics, through which financial investments are forever in search of new opportunities for profit.[65] The most important subjectivity driving green growth is that of the competitive entrepreneur, whether individuals, firms or states: for Mazzucato the ability of states 'to play an *entrepreneurial* role in society' – especially in investing in green technology – is key to 'begin the green industrial revolution and to tackle climate change'.[66]

Most national development strategies in the global South emphasise the importance of achieving higher levels of growth and development, and many present green technologies and investments as one way to 'leapfrog' older and more inefficient industrialisation paths.[67] Ethiopia, discussed above, also illustrates a strongly growth-orientated vision of the green economy. China is perhaps the most obvious example, however, of a country which is aggressively pursuing an expansionist growth path, but which is also investing heavily in new hi-tech green sectors such as wind energy and solar power.[68] Harnessing natural resources – such as water – more efficiently is also a crucial element of China's development plans, and dealing with the impacts of existing dirty industrialisation (most notably air pollution) is a key plank of China's engagement with the green economy.[69] South Africa is another high-profile example of a developing country which has enthusiastically bought into the green growth discourse, and in which sectors like renewable energy are seen as a potential driver of development and jobs: indeed, green growth is a discourse evident across the global South as well as in the industrialised North.[70]

India's vision of economic progress and development is another important example of the discourse of green growth, and the historic experience of agricultural modernisation and the 'green revolution' has shaped the country's current engagement with environmental discourses.[71] Assisted by international donors and agencies, including the Ford Foundation, agricultural development projects to promote industrial farming, mechanisation, irrigation, fertilisers and pesticides, as well as improved crops, were promoted in India from the 1960s onwards, and these all played a major role in rural development.[72] India's green revolution (which is different from the way in which the green revolution discourse is defined below) is also an excellent example of the new problems which can be caused by the enthusiastic adoption of new technologies and practices: the pollution of waterways and aquifers through increased levels of pesticide and fertiliser use has been one legacy, for example.[73] Current debates over greening the economy in India have become characterised by a 'growth versus the environment' framing, with public criticism of the environment minister for opposing the fast-tracking of approval for major dam and energy construction projects, and others suggesting that inclusive growth rather than green growth should be India's primary focus.[74] Despite this, continued agricultural-driven growth is an important strategic aim for Indian politicians. In the July 2014 budget speech Indian Finance Minister Arun Jaitley declared an aim of 'sustaining a growth of 4% in Agriculture', fuelled by a 'technology driven second green revolution with focus on higher productivity'.[75] While such a development path clearly offers huge benefits and profits for some, it is doubtful whether it can also deliver poverty reduction, greater social and economic equality and environmental sustainability.[76]

Whatever the implications, green economy strategies in India clearly involve important shifts in the relationship between the state and society, and between the global North and global South. As in Arun Agrawal's classic study of the protection of green subjectivities through local forest management in India, new national green economy strategies could be seen as part of what he terms 'environmentality': 'the knowledges, politics, institutions and subjectivities that come to be linked together with the emergence of the environment as a domain that requires regulation and protection' – but with the addition that the environment is now also a driver of growth as well as a domain requiring protection.[77] Green growth means overcoming resource scarcity and ensuring energy security, as well as providing potential new sources of external development funding at a point when many countries are cutting development aid to India. India has been one of the largest beneficiaries of Clean Development Mechanism projects outside China and, as one of the remaining outspoken critics of new binding emission reduction targets in the climate negotiations, the country may stand to be a key beneficiary of new funds made available through the Green Climate Fund. However, the key determinant of the direction of India's green economy will be government policy. In 2014 the new Energy Minister, Piyush Goyal, declared that India would become a 'renewables superpower', with $100 billion (£62 billion) to be invested in renewable energy in the next five years and green energy corridors to promote investment in renewables set-up across the country. The previous government set a solar target of 20GW by 2022, but Goyal said in an interview with the *Guardian* that this will be smashed: 'It will be much, much larger. I think for India to add 10GW a year [of solar] and six, or seven or eight of wind every year is not very difficult to envisage'.[78]

Green transformation

The third key discourse of the green economy is green transformation. This is closest to the older discourse of sustainable development, and particularly the Brundtland Report's vision of sustainable development as a realignment of prevailing growth models and development paths.[79] UNEP declares bluntly that we cannot miss this 'chance to fundamentally shift the trajectory of human civilization'.[80] The scope of visibility of this discourse is therefore planetary and civilisational – yet there is also a reliance on existing actors and structures. Thus economic growth remains the driver of progress, the environment is a resource for human development and green developmental states are the regulators and guarantors of development.[81]

Typical technologies and forms of knowledge include Keynesian strategies of public investments and fiscal stimulus, mobilised for 'green' ends: clean air, water and food, safe and efficient public transport, tree-planting campaigns, and so on.[82] UNEP's 'Global Green New Deal' invokes the historical precedent of Franklin D Roosevelt's New Deal in 1930s America, and would include green stimulus policies such as 'support for renewable energy, carbon capture and sequestration, energy efficiency, public transport and rail, and improving electrical grid transmission'.[83] The International Labour Organization (ILO) has suggested that 'a greener economy could lead to net gains of up to 60 million jobs'.[84] South Africa's 'Green Economy Accord', Mozambique's 'Green Economy Action Plan' and Rwanda's 'National Strategy for Climate Change and Low Carbon Development' are all examples of national strategies with strong elements of this transformational discourse.[85]

The most prominent and most frequently cited example of green transformation anywhere in the world (let alone just in the global South) is South Korea.[86] According to UNEP, 'in January 2009, the Government of the Republic of Korea responded to the deepening recession with a Green New Deal, a fiscal stimulus package equivalent to US\$ 38.1 billion of which 80% was allocated to more efficient use of resources such as fresh-water, waste, energy-efficient buildings, renewable energies, low-carbon vehicles, and the rail network'.[87] This package of green spending was 95% of the fiscal stimulus spent in response to the financial crisis, 1.2% of GDP, and it aimed to directly create at least 334,000 new jobs.[88] Longer-term plans to double the generation of energy from renewable sources and enhance resource and material efficiency hope to create 1.56–1.81 million jobs in green industries, and to contribute to a substantial reduction in greenhouse gases.[89] Korea's spending dwarfed all other green economy commitments in response to the financial crisis, and has accordingly been held up as a notable example of a country trying to shift its development path from a previous reliance on fossil fuels and dirty manufacturing, to a more modern, greener economy.

There are clear similarities here to the 'green growth' discourse, and indeed Korea is also held up by the OECD as being 'at the forefront of green growth initiatives'.[90] The Global Green Growth Institute (GGGI) is based in Seoul, and exists to accelerate 'the transition toward a new model of economic growth – green growth – founded on principles of social inclusivity and environmental sustainability'.[91] It lists its members as Brazil, Cambodia, Columbia, China, Ethiopia, India, Indonesia, Kazakhstan, Peru, Mexico, Mongolia, Morocco, The Philippines, Rwanda, South Africa, Thailand, United Arab Emirates and Vietnam.[92] The GGGI dates its emergence to President Lee Myung-bak's proclamation of a 'Low Carbon, Green Growth' development vision in 2008, which 'aimed to shift the current development paradigm of quantity-oriented, fossil-fuel dependent growth to quality-oriented growth with an emphasis on the use of new and renewable energy resources'.[93] Korea's 'National Strategy for Green Growth' (2009–50) also explicitly invokes the green growth discourse.[94] However, discursively, there is a difference between green growth and green transformation. Whereas the former tends to present growth as an end rather than a means, green transformation discourses call for the model of growth itself to be transformed, and involve explicitly political interventions into transforming the structure of the economy.

Of course, grand plans to transform economic infrastructure are always political and contested, and Korea's transformation has attracted criticism from a range of directions: those who fear it is going too fast, or not fast enough; those whose livelihoods and housing are threatened by new industries or the decline of old industries; and opponents of the hubris and risk of the whole endeavour. Tienhaara notes that 'activists in South Korea have strongly opposed the use of a substantial amount of stimulus funds on a controversial project to dam and dredge four major rivers, putting a number of endangered species at serious risk'.[95] Others have praised Korea's ambition as an example of the state putting transformative visions of green urbanism into practice: for Mark Swilling Seoul is one of the best examples of 'extraordinary city-wide partnerships to completely reinvent the city from a sustainability perspective...where the highway through the city was replaced with the river that used to be there'.[96] According to Reuters, 'both rich and poor nations are turning to Seoul for lessons in green-powered development', and hoping to '"leapfrog" over the dirty technologies like coal that were ushered in by the first industrial revolution'.[97] To be clear: the argument here is not that Korea's development strategy is convincingly progressive, ecologically sustainable and transformational. However, there is greater discursive emphasis in its articulation of the green economy on these features than in the discourses evident in other leading examples, such as India and China, as well as in much of the global North. This discursive framing creates both opportunities and new potential for environmental activists to hold Korean politicians to their promises.

Green revolution

Environmental activists have therefore been crucial in building pressure for more transformational green economy strategies, but many such activists are now publicly sceptical of mainstream discourses of green growth and green transformation.[98] As such they are often attached to a discourse of 'green revolution' in the sense used by many environmentalists in the 1960s and 1970s: a radical realignment of economic (and hence social and political) relationships to bring them in line with natural limits and ecological virtues.[99] The range of vision opened up by deep ecologists, eco-socialists, eco-feminists, indigenous peoples and others includes the natural world, a broader conception of human society and an interconnected cosmos.[100] The 'limits to growth' proclaimed by environmentalists in the 1970s have inspired a range of alternative forms of knowledge such as de-growth, steady-state economics or prosperity without growth.[101] The idea that the economic system requires 'greening' in order to resolve contradictions and end the systematic exploitation of nature has been a central tenet of these traditions.

Unsurprisingly this radical vision has often been marginalised in mainstream discussions of the green economy, and a few years ago it was hard to see examples of such a discourse in practice outside small communal 'eco-villages' and retreats. However, radicals have been reinvigorated by the example of Latin American states, led by Bolivia, which have constitutionally recognised the Rights of Mother Earth, and whose aim is *buen vivir* rather than endless economic growth.[102] Costa Rica has received international praise for its revolutionary approach to the green economy, including turning payments for ecosystem services from a principle into practice and its commitment to eco-tourism, and it comes at or towards the top of international indices such as the Happy Planet Index and the Global Green Economy Index.[103] Yet even the champions of the '*buen vivir*' discourse have attracted criticism for failing to live-up to their revolutionary potential in practice. Costa Rica has discovered that rising land values mean that it has become 'un-economic' not to farm land, even when ecosystem services are fully priced.[104] Bolivia has continued to promote extractive mining projects, despite recognising the rights of Mother Earth in its constitution.[105] Ecuador's ambitious plan to protect the Yasuní National Park failed in 2013 when international support to offset the sacrificed oil revenues did not materialise.[106]

Brazil is another example of a national development trajectory that, while sometimes progressive, hardly lives up to the potential for a green revolution. Brazil has received praise for its green jobs strategy, investment in renewable energy and ambitious climate targets.[107] It has sought to halt the net loss of forest coverage by 2015, and is committed to reducing between 36.1% and 38.9% of its projected greenhouse gas emissions by 2020. It has also included green jobs as a key element in its national development policy, supported by the ILO's *Programa empregos verdes*.[108] The *Bolsa Verde* (Green Grant) programme was launched in 2011 and provides income grants to families who live in extreme poverty in rural areas in exchange for maintenance and sustainable use of natural resources, while the *Bolsa Floresta* programme provides forest households with monthly payments into credit-card accounts for practising 'farming without fire' (which is monitored by satellite).[109] Brazil is also a key international player in negotiations on deforestation and climate change targets.[110]

However, despite the mainstream dominance of green growth and green transformation in Brazil, it is also an important site of significant manifestations of the green revolution discourse. The land occupations and demands for food sovereignty by social movements like the Landless Workers' Movement (MST) and coalitions like La Vía Campesina have been hailed by activists and environmentalists worldwide as one of the best candidates for a mass-based, radical green movement.[111] As well as contesting the discursive framings of green growth, the commodification of ecosystem services and narratives of agricultural modernisation, movements like MST have carried out hundreds of thousands of illegal land occupations since the 1980s, thereby enabling poor communities to access land and return to land they have been evicted from.[112] For Philip McMichael, 'transnational movements such as Vía Campesina advocate a world to gain – a world beyond the catastrophe of the corporate market regime, in which agrarianism is re-valued as central to social and ecological sustainability'.[113] It is through such movements – both their autonomous and prefigurative attempts to build new societies, but also their pressure on transformative states to become green states – that it may be possible to build greener coalitions for radical and indeed revolutionary change.

Conclusion

The examples profiled here are far from exhaustive, nor do they fit entirely neatly into the categories assigned. The discourse of green growth can be identified in all the national strategies examined, for example. In fact, all four discourses have elements present in all of the actual empirical manifestations of the green economy, in the global South as well as the global North. However, it is of analytical and heuristic value to disaggregate the politics of the green economy and highlight some of the discursive tensions that lie behind the disagreements over what the green economy means.

The major commonality in all these national strategies and articulations of the green economy is the central role of the state. The creation of markets in carbon credits and ecosystem services, and the stimulation of economies through agricultural investment and employment programmes, is one of the central ways in which state institutions are seeking to re-legitimate their regulatory or steering role in local, national and global economies. Renewed demands for 'entrepreneurial' or 'developmental' states which can forge new green economy pathways are being heard from policy makers, activists and academics.[114] As UNEP's Global Green New Deal makes clear, the green economy is necessary precisely 'because the unregulated market cannot resurrect itself on its own from a failure of a historical proportion without significant and coordinated government interventions'.[115] Similarly the UNCTAD Economic Development in Africa Report 2012 argues that 'a major negative side effect of the structural adjustment phase was the erosion of State capacities. Building up developmental States' capabilities to formulate and implement structural transformation policies will thus be an important challenge.'[116] Even discourses of green revolution cannot neglect or bypass the state: in Bolivia and Ecuador they have worked through state structures and constitutional process, and radical social movements like La Vía Campesina have sought recognition, concessions and reforms from state institutions and political actors. None of these interventions is neutral or a win-win solution: they all involve winners and losers, and those best positioned to take advantage in the medium to long term from these green economy and green growth programmes are existing elites.

For this reason we should be wary not only of discourses of the green economy, but also of the emergence of green states in the global South. Green states – defined here as states which use the discourses of environmentalism and 'green branding' to legitimate their development politics – might have more in common with colonial, racist or authoritarian states than some of the more optimistic narratives of ecological modernisation assume. For Robyn Eckersley, 'a proliferation of transnationally oriented green states' is 'likely to provide a surer path to a greener world'.[117] Yet the green growth discourse, currently dominant in the global South, threatens to continue to exacerbate existing highly inequitable and ecologically damaging growth patterns, and to further discredit environmental discourses in the process by legitimating big infrastructural projects like dams and intensive commercial agriculture. Green states – and green social movements – are surely necessary in forging greener economies than those we currently have, but more radical environmentalists will seek to ensure that the green economy under construction is transformative or even revolutionary, rather than a defence of the status quo of growth at any cost.

Acknowledgements

A version of this article was originally presented at the 'Green Economy in the South' conference, University of Dodoma, Tanzania, 8–10 July 2014. Many thanks to the journal editors and three anonymous reviewers for their constructive comments.

Notes

1. UNEP, *Towards a Green Economy*, 16.
2. Brockington, "A Radically Conservative Vision?"; Goodman and Salleh, "The 'Green Economy'"; Tienhaara, "Varieties of Green Capitalism"; and Wanner, "The New 'Passive Revolution'."
3. Death, *Governing Sustainable Development*, 14.
4. Benson et al., *Green Economy Barometer*, 4; and Mol and Sonnenfeld, "Ecological Modernisation."
5. Bäckstrand and Kronsell, *Rethinking the Green State*; Dryzek et al., *Green States and Social Movements*; and Eckersley, *The Green State*.
6. Death, "Green States in Africa."
7. See https://greeneconomyinthesouth.wordpress.com/. Typical examples include Brockington et al., *Nature Unbound*; Büscher and Fletcher, "Accumulation by Conservation"; and Nelson, *Community Rights*.
8. Peet et al., *Global Political Ecology*.
9. For example, Resnick et al., *The Political Economy of Green Growth*.
10. Konings, "Renewing State Theory," 174. See also Mazzucato, *The Entrepreneurial State*.
11. Death, "Green States in Africa."
12. UNEP, *Towards a Green Economy*, 16.
13. Meadowcroft, "Greening the State?," 69.
14. World Bank, *Inclusive Green Growth*, 2.
15. Pearce et al., *Blueprint for a Green Economy*.
16. Barbier, "The Policy Challenges," 234.
17. Jessop, "Economic and Ecological Crises"; and Scoones et al., "The Politics of Green Transformations," 1.
18. Stern, *The Economics of Climate Change*.
19. Tienhaara, "Varieties of Green Capitalism."
20. UNEP, *Global Green New Deal*, 1.
21. UNEP, *Towards a Green Economy*, 14.
22. Ibid., 15.
23. Mazzucato, *The Entrepreneurial State*; Scoones et al., "The Politics of Green Transformations"; and UNEP, *Global Green New Deal*.
24. Clapp and Dauvergne, *Paths to a Green World*; and Kim and Thurbon, "Developmental Environmentalism."
25. Harman and Williams, "International Development in Transition."
26. Mol and Buttel, "The Environmental State under Pressure," 2.
27. Williams, "Rethinking the Developmental State."
28. Dryzek et al., *Green States and Social Movements*; and Eckersley, *The Green State*. See also Death, "Green States in Africa."
29. Migdal, *Strong Societies and Weak States*, 4.
30. Williams, "Rethinking the Developmental State," 23.
31. See Death, *Governing Sustainable Development*, chap. 2.
32. Doty, *Imperial Encounters*; and Mitchell, *Rule of Experts*.
33. Doty, *Imperial Encounters*, 5–6.
34. Foucault, *The Archaeology of Knowledge*, 54.
35. Death, *Governing Sustainable Development*.
36. Dryzek, *The Politics of the Earth*.
37. A fuller account of the methodology of this approach to discourses of government can be found in Death, *Governing Sustainable Development*, 20.
38. Benson et al., *Green Economy Barometer*; UNDP, *Human Development Report 2011*; UNEP, *Global Green New Deal*; UNEP, *Towards a Green Economy*; and World Bank, *Inclusive Green Growth*.
39. Tienhaara, "Varieties of Green Capitalism."
40. Ferguson, "The Green Economy Agenda," 27. See also Bina, "The Green Economy"; and Scoones et al., "The Politics of Green Transformations," 9.
41. This builds upon and extends the argument in Death, "The Green Economy in South Africa.".

42. UNDP, *Human Development Report 2011*.
43. UNEP, *Towards a Green Economy*, 40.
44. World Bank, *Inclusive Green Growth*, 2.
45. World Bank, "World Bank Climate Change Strategy."
46. Methmann and Oels, "Vulnerability."
47. World Bank, *Inclusive Green Growth*.
48. UNDP, *Human Development Report 2011*, 5.
49. Ibid., 78.
50. Federal Democratic Republic of Ethiopia, *Ethiopia's Climate-resilient Green Economy*, 1. See also Benson et al., *Green Economy Barometer*.
51. Lautze and Maxwell, "Why do Famines Persist?"
52. World Bank, "Ethiopia Economic Update."
53. Federal Democratic Republic of Ethiopia, *Ethiopia's Climate-resilient Green Economy*, 81.
54. Jones and Carabine, *Exploring Political and Socio-economic Drivers*.
55. UNEP, *Towards a Green Economy*, 20.
56. Petersen, "Developing Climate Adaptation."
57. Ibid., 562; ILO, *Sustainable Development*, 66; and UNDP, *Human Development Report 2011*, 78.
58. Petersen, "Developing Climate Adaptation," 575.
59. Ibid.
60. Adem, "The Local Politics"; and Lefort, "Free Market Economy."
61. Brassett et al., "Introduction"; Jones and Carabine, *Exploring Political and Socio-economic Drivers*, 2; and Methmann and Oels, "Vulnerability," 283–284.
62. World Bank, *Inclusive Green Growth*, 3.
63. Newell and Paterson, *Climate Capitalism*, chap. 4.
64. Büscher and Fletcher, "Accumulation by Conservation," 11.
65. Benson et al., *Green Economy Barometer*, 3; and World Bank, *Inclusive Green Growth*, 2.
66. Mazzucato, *The Entrepreneurial State*, 1, 118. Emphasis in original.
67. Benson et al., *Green Economy Barometer*; and UNEP, *Towards a Green Economy*, 3.
68. Lewis, *Green Innovation in China*.
69. UNEP, *China's Green Long March*; and Weng et al., *China's Path to a Green Economy*.
70. Death, "The Green Economy in South Africa."
71. UNEP, *Towards a Green Economy*, 11.
72. Agrawal, *Environmentality*.
73. Resnick et al., "The Political Economy of Green Growth," 222.
74. Dubash, "Toward Enabling and Inclusive Global Environmental Governance," 50; and Gupta, "Why India's Green Growth."
75. Jaitley, "Budget Speech."
76. Shiva, *Making Peace with the Earth*. See also UNDP, *Human Development Report 2011*, 2.
77. Agrawal, *Environmentality*, 226.
78. Carrington, "India will be Renewables Superpower."
79. Dryzek, *The Politics of the Earth*, 147–151; and Death, *Governing Sustainable Development*.
80. UNEP, *Global Green New Deal*, 4.
81. Ibid; UNEP, *Towards a Green Economy*; and Williams, "Rethinking the Developmental State."
82. Newell and Paterson, *Climate Capitalism*, 179; and UNCTAD, *Economic Development in Africa*.
83. Barbier, "Global Governance," 3; and Tienhaara, "Varieties of Green Capitalism," 190.
84. ILO, *Sustainable Development*, xiv.
85. Death, "Green States in Africa."
86. Barbier, "Global Governance"; Brockington, "A Radically Conservative Vision?"; Kim and Thurbon, "Developmental Environmentalism"; Mazzucato, *The Entrepreneurial State*, 121; and Tienhaara, "Varieties of Green Capitalism."
87. UNEP, "Korea's Pathway."
88. Barbier, "Global Governance," 7.
89. UNEP, "Korea's Pathway."
90. OECD, "Green Growth in Action."
91. See Green Growth Institute. Accessed January 26, 2015, http://gggi.org/about-gggi/background/organizational-overview/.
92. See http://gggi.org/activities/ggpi/summaries-by-country/.
93. Green Growth Institute. Accessed January 26, 2015, http://gggi.org/about-gggi/background/organizational-overview/.
94. Kim and Thurbon, "Developmental Environmentalism."
95. Tienhaara, "Varieties of Green Capitalism," 5. See also Kim and Thurbon, "Developmental Environmentalism," 231.
96. Swilling, "Reconceptualising Urbanism," 88.
97. Feldman, "Green Growth."
98. Benson et al., *Green Economy Barometer*, 5.

99. Bookchin, *Toward an Ecological Society*; and Schumacher, *Small is Beautiful.*
100. Dryzek, *The Politics of the Earth*, chaps. 9, 10; Goodman and Salleh, "The 'Green Economy'"; and Shiva, *Making Peace with the Earth.*
101. Bina, "The Green Economy," 1037; Clapp and Dauvergne, *Paths to a Green World*, 9–11; and Jessop, "Economic and Ecological Crises," 23.
102. Goodman and Salleh, "The 'Green Economy'"; and Stevenson, "Representing Green Radicalism."
103. Happy Planet Index. Accessed January 26, 2015, http://www.happyplanetindex.org/data/;Dual Citizen, *Global Green Economy Index*; and UNDP, *Human Development Report 2011*, 4.
104. Porras, "Costa Rica's 'Green Economy'."
105. Fidler, "Bolivia leads Climate Change Fight."
106. Martin and Scholz, "Ecuador's Yasuní–ITT Initiative."
107. UNEP, *Towards a Green Economy*, 15, 19. See also http://gggi.org/brazil-green-growth-planning/, accessed May 19, 2015.
108. Gaetani et al., "Brazil and the Green Economy."
109. ILO, *Sustainable Development.*
110. Allan and Dauvergne, "The Global South."
111. Steward et al., *Towards a Green Food System.*
112. Routledge and Cumbers, *Global Justice Networks.*
113. McMichael, "Peasants make their own History," 210.
114. Kim and Thurbon, "Developmental Environmentalism," 221; Mazzucato, *The Entrepreneurial State*; Meadowcroft, "Greening the State?"; and Williams, "Rethinking the Developmental State."
115. UNEP, *Global Green New Deal*, 4.
116. UNCTAD, *Economic Development in Africa*, 86.
117. Eckersley, *The Green State*, 202.

Bibliography

Adem, Teferi Abate. "The Local Politics of Ethiopia's Green Revolution in South Wollo." *African Studies Review* 55, no. 3 (2012): 81–102.
Agrawal, Arun. *Environmentality: Technologies of Government and the Making of Subjects.* Durham, NC: Duke University Press, 2005.
Allan, Jen Iris, and Peter Dauvergne. "The Global South in Environmental Negotiations: The Politics of Coalitions in REDD+." *Third World Quarterly* 34, no. 8 (2013): 1307–1322.
Bäckstrand, Karin, and Annica Kronsell, eds. *Rethinking the Green State: Environmental Governance towards Climate and Sustainability Transitions.* London: Earthscan, 2015.
Barbier, Edward B. "Global Governance: The G20 and a Global Green New Deal." *Economics: The e-journal* 4 (2010): 1–32. See http://www.economics-ejournal.org/economics/journalarticles/2010-2.
Barbier, Edward B. "The Policy Challenges for Green Economy and Sustainable Economic Development." *Natural Resources Forum* 35 (2011): 233–245.
Benson, Emily, Steve Bass, and Oliver Greenfield. *Green Economy Barometer: Who is doing What Where, and Why?* London: Green Economy Coalition, February 2014.
Bina, Olivia. "The Green Economy and Sustainable Development: An Uneasy Balance?" *Environment and Planning C: Government and Policy* 31 (2013): 1023–1047.
Bookchin, Murray. *Toward an Ecological Society.* Montreal: Black Rose Books, 1980.
Brassett, James, Stuart Croft, and Nick Vaughan-Williams. "Introduction: An Agenda for Resilience Research in Politics and International Relations." *Politics* 33, no. 4 (2013): 221–228.
Brockington, Dan. "A Radically Conservative Vision? The Challenge of UNEP's Towards a Green Economy." *Development and Change* 43, no. 1 (2012): 409–422.
Brockington, Dan, Rosaleen Duffy, and Jim Igoe. *Nature Unbound: Conservation, Capitalism and the Future of Protected Areas.* London: Earthscan, 2008.
Büscher, Bram, and Robert Fletcher. "Accumulation by Conservation." *New Political Economy* 20, no. 2 (2015): 273–298.
Carrington, Damian. 2014. "India will be Renewables Superpower, says Energy Minister." *Guardian*, October 1.
Clapp, Jennifer, and Peter Dauvergne. *Paths to a Green World: The Political Economy of the Global Environment.* Cambridge, MA: MIT Press, 2005.
Death, Carl. *Governing Sustainable Development: Partnerships, Protests and Power at the World Summit.* Abingdon: Routledge, 2010.
Death, Carl. "Green States in Africa: Beyond the Usual Suspects." *Environmental Politics* 25, no. 1 (2016).
Death, Carl. "The Green Economy in South Africa: Global Discourses and Local Politics." *Politikon: South African Journal of Political Studies* 41, no. 1 (2014): 1–22.
Doty, Roxanne Lynne. *Imperial Encounters: The Politics of Representation in North–South Relations.* Minneapolis: University of Minnesota Press, 1996.
Dryzek, John S. *The Politics of the Earth: Environmental Discourses.* Oxford: Oxford University Press, 2012.

Dryzek, John S., David Downes, Christian Hunold, David Schlosberg, and Hans-Kristian Hernes. *Green States and Social Movements: Environmentalism in the United States, United Kingdom, Germany, and Norway.* Oxford: Oxford University Press, 2003.

Dual Citizen. *Global Green Economy Index 2014: Measuring National Performance in the Green Economy.* Washington, DC: Dual Citizen, 2014.

Dubash, Navros K. "Toward Enabling and Inclusive Global Environmental Governance." *Journal of Environment and Development* 21, no. 1 (2012): 48–51.

Eckersley, Robyn. *The Green State: Rethinking Democracy and Sovereignty.* Cambridge, MA: MIT Press, 2004.

Federal Democratic Republic of Ethiopia. *Ethiopia's Climate-resilient Green Economy.* Addis Ababa: Federal Democratic Republic of Ethiopia, 2011.

Feldman, Stacey. "Green Growth, South Korea's National Policy, Gaining Global Attention." Reuters, January 26, 2011.

Ferguson, Peter. "The Green Economy Agenda: Business as Usual or Transformational Discourse?" *Environmental Politics* 24, no. 1 (2015): 17–37.

Fidler, Richard. "Bolivia leads Climate Change Fight." *Green Left Weekly*, October 20, 2014.

Foucault, Michel. *The Archaeology of Knowledge.* London: Routledge, 2002.

Gaetani, Francisco, Ernani Kuhn, and Renato Rosenberg. "Brazil and the Green Economy: A Panorama." *Política Ambiental* (June 2011): 76–85.

Goodman, James, and Ariel Salleh. "The 'Green Economy': Class Hegemony and Counter-Hegemony." *Globalizations* 10, no. 3 (2013): 411–424.

Gupta, Alok. "Why India's Green Growth Dream is turning into a Nightmare." Green Economy Coalition, December 2012. http://www.greeneconomycoalition.org/know-how/why-india%E2%80%99s-green-growth-dream-turning-nightmare.

Harman, Sophie, and David Williams. "International Development in Transition." *International Affairs* 90, no. 4 (2014): 925–941.

ILO. *Sustainable Development, Decent Work and Green Jobs.* Geneva: ILO, 2013.

Jaitley, Arun. "Budget Speech, Indian Minister of Finance, 2014–5." July 10, 2014. http://pib.nic.in/archieve/others/2014/jul/gbEngSpeech.pdf.

Jessop, Bob. "Economic and Ecological Crises: Green New Deals and No-growth Economies." *Development* 55, no. 1 (2012): 17–24.

Jones, Lindsey, and Elizabeth Carabine. *Exploring Political and Socio-economic Drivers of Transformational Climate Policy: Early Insights from the Design of Ethiopia's Climate Resilient Green Economy Strategy.* London: ODI, 2013.

Kim, Sung-Young, and Elizabeth Thurbon. "Developmental Environmentalism: Explaining South Korea's Ambitious Pursuit of Green Growth." *Politics & Society* 43, no. 2 (2015): 213–240.

Konings, Martijn. "Renewing State Theory." *Politics* 30, no. 3 (2010): 174–182.

Lautze, Sue, and Daniel Maxwell. "Why do Famines Persist in the Horn of Africa? Ethiopia, 1999–2003." In *The New Famines: Why Famines Persist in an Era of Globalization*, edited by Stephen Devereux, 222–244. Abingdon: Routledge, 2007.

Lefort, René. "Free Market Economy, 'Developmental State' and Party-State Hegemony in Ethiopia: The Case of the 'Model Farmers'." *Journal of Modern African Studies* 50, no. 4 (2012): 681–706.

Lewis, Joanna I. *Green Innovation in China: China's Wind Power Industry and the Global Transition to a Low-carbon Economy.* New York: Columbia University Press, 2013.

Martin, Pamela L., and Imme Scholz. "Ecuador's Yasuní–ITT Initiative: What can we learn from its Failure?" *International Development Policy* 5 (2014): 1–11.

Mazzucato, Mariana. *The Entrepreneurial State: Debunking Public vs. Private Sector Myths.* London: Anthem Press, 2014.

McMichael, Philip. "Peasants make their own History, but not just as they Please..." *Journal of Agrarian Change* 8, nos. 2–3 (2008): 205–228.

Meadowcroft, James. "Greening the State?" In *Comparative Environmental Politics: Theory, Practice, and Prospects*, edited by Paul F. Steinberg and Stacy D. VanDeveer, 63–87. Cambridge, MA: MIT Press, 2012.

Methmann, Chris, and Angela Oels. "Vulnerability." In *Critical Environmental Politics*, edited by Carl Death, 277–286. Abingdon: Routledge, 2014.

Migdal, Joel S. *Strong Societies, and Weak States: State-Society Relations and State Capabilities in the Third World.* Princeton, NJ: Princeton University Press, 1988.

Mitchell, Timothy. *Rule of Experts: Egypt, Techno-politics, Modernity.* Berkley: University of California Press, 2002.

Mol, Arthur P. J., and Frederick H. Buttel. "The Environmental State under Pressure: An Introduction." In *The Environmental State under Pressure*, edited by Arthur P. J. Mol and Frederick H. Buttel, 1–12. Oxford: Elsevier, 2002.

Mol, Arthur P. J., and David A. Sonnenfeld. "Ecological Modernisation around the World: An Introduction." *Environmental Politics* 9, no. 1 (2000): 1–14.

Nelson, Fred, ed. *Community Rights, Conservation and Contested Land: The Politics of Natural Resource Governance in Africa.* Abingdon: Earthscan, 2010.

Newell, Peter, and Matthew Paterson. *Climate Capitalism: Global Warming and Transformation of the Global Economy.* Cambridge: Cambridge University Press, 2010.

OECD. "Green Growth in Action: Korea." Accessed December 22, 2014. http://www.oecd.org/korea/green growthinactionkorea.htm.

Pearce, David, Anil Markandya, and Edward Barbier. *Blueprint for a Green Economy.* London: Earthscan, 1989.

Peet, Richard, Paul Robbins, and Michael J. Watts, eds. *Global Political Ecology.* Abingdon: Routledge, 2011.

Petersen, Nicole D. "Developing Climate Adaptation: The Intersection of Climate Research and Development Programmes in Index Insurance." *Development and Change* 43, no. 2 (2012): 557–584.

Porras, Ina. "Costa Rica's 'Green Economy' shows that Money can grow on Trees." *Guardian*, June 29, 2012.

Resnick, Danielle, Finn Tarp, and James Thurlow. "The Political Economy of Green Growth: Cases from Southern Africa." *Public Administration and Development* 32 (2012): 215–228.

Routledge, Paul, and Andrew Cumbers. *Global Justice Networks: Geographies of Transnational Solidarity.* Manchester, NH: Manchester University Press, 2009.

Schumacher, Ernst F. *Small is Beautiful: A Study of Economics as if People Mattered.* New York: Harper and Row, 1973.

Scoones, Ian, Peter Newell, and Melissa Leach. "The Politics of Green Transformations." In *The Politics of Green Transformations*, edited by Ian Scoones, Melissa Leach and Peter Newell, 1–24. Abingdon: Routledge, 2015.

Shiva, Vandana. *Making Peace with the Earth.* London: Pluto Press, 2012.

Stern, Nicholas. *The Economics of Climate Change: The Stern Review.* Cambridge: Cambridge University Press, 2007.

Stevenson, Hayley. "Representing Green Radicalism: The Limits of Statebased Representation in Global Climate Governance." *Review of International Studies* 40, no. 1 (2014): 177–201.

Steward, Corrina, Maria Aguiar, Nikhil Aziz, Jonathan Leaning, and Daniel Moss. *Towards a Green Food System: How Food Sovereignty can Save the Environment and Feed the World.* Boston, MA: Grassroots International, 2008.

Swilling, Mark. "Reconceptualising Urbanism, Ecology and Networked Infrastructures." *Social Dynamics: A Journal of African Studies* 37, no. 1 (2011): 78–95.

Tienhaara, Kyla. "Varieties of Green Capitalism: Economy and Environment in the Wake of the Global Financial Crisis." *Environmental Politics* 23, no. 2 (2014): 187–204.

UNCTAD. *Economic Development in Africa Report 2012: Structural Transformation and Sustainable Development in Africa.* New York: United Nations, 2012.

UNDP. *Human Development Report 2011: Sustainability and Equity – A Better Future for All.* New York, NY: UNDP, 2011.

UNEP. *Global Green New Deal.* Nairobi: UNEP, 2009.

UNEP. *Towards a Green Economy: Pathways to Sustainable Development and Poverty Eradication.* Nairobi: UNEP, 2011.

UNEP. *China's Green Long March: A Study of Renewable Energy, Environmental Industry and Cement Sectors.* Nairobi: UNEP, 2013.

UNEP. "Korea's Pathway to a Green Economy." Accessed December 22, 2014. http://www.unep.org/greenecon omy/AdvisoryServices/Korea/tabid/56272/Default.aspx.

Wanner, Thomas. "The New 'Passive Revolution' of the Green Economy and Growth Discourse: Maintaining the 'Sustainable Development' of Neoliberal Capitalism." *New Political Economy* 20, no. 1 (2015): 21–41.

Weng, Xiaoxue, Zhanfeng Dong, Wu Qiong, and Ying Qin. *China's Path to a Green Economy: Decoding China's Green Economy Concepts and Policies.* London: IIED, 2015.

Williams, Michelle. "Rethinking the Developmental State in the Twenty-first Century." In *The End of the Developmental State?*, edited by Michelle Williams, 1–29. Abingdon: Routledge, 2014.

World Bank. "World Bank Climate Change Strategy for Africa calls for Adaptation, Mitigation and Additional Financing." Press release, November 30, 2010.

World Bank. *Inclusive Green Growth: The Pathway to Sustainable Development.* Washington, DC: World Bank, 2012.

World Bank. "Ethiopia Economic Update – Laying the Foundation for Achieving Middle Income Status." Press release, June 18, 2013.

Tourism and the green economy: inspiring or averting change?

Melanie Stroebel

School of Environment, Education and Development, University of Manchester, UK

This paper investigates how tourism stakeholders conceptualise tourism in a green economy and how they foresee the transition to progress. With the meaning of a green economy remaining contested, the political agenda that the concept entails in a particular context can be far from clear. The paper provides a qualitative analysis of *Towards a Green Economy* and the publication *Green Growth and Travelism* to explore the implementation strategies and political agendas of tourism stakeholders. It outlines how stakeholders argue in line with international organisations that tourism can contribute to growth, development and poverty alleviation, while reducing environmental impacts. However, some researchers challenge the foundations of the green growth discourse. An exploration of these contradictions and of the political and economic implications of climate change leads the paper to argue that the particular framing of the green economy presents tourism in a way that sets the industry up for continued growth, while marginalising a much-needed radical transformation.

Introduction

In *Towards a Green Economy* the United Nations Environment Programme (UNEP) identifies tourism as a brown sector that, provided green investment, could contribute to a green economy.[1] Parallel to this publication, the tourism industry has engaged in depth with its role in greening both tourism and associated activities. For example, *Green Growth and Travelism*, the publication analysed in this paper, presents a reflective and aspirational engagement of 46 leading tourism representatives with the subject.[2] Despite, or maybe because of, increasing attention, the political agendas pursued under the banner of a green economy vary greatly. While ongoing work discerns political agendas of a green economy more generally,[3] this paper focuses on a particular industry. It seeks to develop a better understanding of how certain tourism stakeholders present the green economy and envisage the transition in the publications *Green Growth and Travelism* and *Towards a Green Economy*; this permits us to identify the rules and norms of climate change governance that the green economy discourse promotes for the wider tourism industry.

Tourism's role in the transition towards a green economy builds on the understanding that 'green investment in tourism can contribute to economically viable and robust growth, decent work creation, poverty alleviation, improved efficiency in resource use and reduced environmental degradation'.[4] However, not all evaluations of tourism draw such positive conclusions. In particular tourism's contribution to development and its climate change impact have long been areas of concern.[5] By comparing the green economy discourse in *Green Growth and Travelism* and UNEP's *Towards a Green Economy* with peer-reviewed research on development and climate change mitigation, the paper identifies conflicting understandings between practitioners and researchers. Building on these contradictions, the paper argues that the green economy framing is not only materially challenging; more importantly, it constitutes a discursive strategy to secure continued growth, while marginalising a necessary radical transformation of the industry in order to effectively reduce its contribution to climate change.

Different notions of a green economy

It is the dilemma of all-encompassing discourses like the green economy that they also provide space for vastly different political agendas. While Pearce's *Blueprint for a Green Economy* aimed to protect the environment by uniting economy and environment through valuation and accounting, recent elaborations have become much broader, attempting at times a reorientation of capitalism.[6] UNEP broadly defines a green economy as one that leads to 'improved human well-being and social equity, while significantly reducing environmental risks and ecological scarcities'.[7] *Towards a Green Economy* states: 'growth in income and employment are driven by public and private investments that reduce carbon emissions and pollution, enhance energy and resource efficiency, and prevent the loss of biodiversity and ecosystem service'.[8] What precisely constitutes a green economy and whether and for whom it is desirable can be unclear and requires close scrutiny, as different framings promote 'a shared way of apprehending the world' that can discursively influence global environmental politics.[9]

Political agendas of a green economy

The sought-after outcomes of a green economy vary notably. For Bowen and Frankhauser a green economy is a strategy to decarbonise growth;[10] for UNEP it provides the path to sustainable development.[11] Cook and Smith understand it as a shift away from the social to the environmental and economic dimensions of sustainability.[12] Tienhaara and also Brand argue that it barely differs from sustainable development.[13] 'Green economy' is clearly a contentious term that serves a diverse range of political agendas.

The diversity of political agendas becomes clear in efforts to classify different green economy understandings. Ehresman and Okereke distinguish a 'thin green economy' (one that promotes growth to alleviate poverty, reduce inequality and solve environmental issues) from a 'thick green economy' (one that assumes ecological limits and, therefore, for the poor to develop requires restrictions and possibly de-growth in developed nations) and from a 'moderate green economy' (one that accepts the market-based approach, but sees a need for reform to reduce the gap between rich and poor).[14] Similarly Death's analysis of the green economy discourse in South Africa identifies four different notions: green revolution, green transformation, green growth and green resilience.[15] While a 'green revolution' aims to radically transform the economy–nature relationship by removing the dominant growth paradigm, a 'green transformation' maintains the basic elements and assumptions of the current economic model; green investments and financial incentives are intended to shift the economy towards increased social justice and equity. 'Green resilience' is a more cautious and protective strategy, focused on adapting the economy to cope with changes. 'Green growth', the dominant discourse in the South African context, sees opportunities for growth in new products, markets and consumption patterns. These divergent understandings of a green economy come as a result of distinct assumptions and perceptions pertaining to the role of growth and its limits, to technological innovation and to the role of the market.

Foundations and implications of diverging political agendas

While some researchers expect economic growth to be the outcome of a green economy, others view green economic growth as the vehicle towards development and poverty alleviation.[16] Growth-centred perspectives often place great importance on technological innovation to decouple environmental impacts from growth. However, they may overlook the fact that rebound effects may increase overall environmental impacts, as improved environmental performance may increase consumption.[17] An entirely different policy direction becomes essential, if one views the earth's resources as finite and consequently sees limits to growth.[18] Such an understanding lends itself to the climate change context, with its, albeit socially constructed, dangerous limit to CO_2 concentration in the atmosphere. It calls for more radical and transformative approaches to growth, such as a steady-state economy and (sustainable) de-growth.[19] Kosoy et al, who challenge the growth paradigm, argue that the excessive consumption patterns that have contributed to environmental crises need to be addressed to ensure resource use is equitable and within ecological limits.[20] Sustainable development consequently entails reconsidering lifestyles,[21] and requires regulatory frameworks to ensure production and consumption return to sustainable levels.[22] It needs to be noted that a contracting economy is an alternative that UNEP's *Towards a Green Economy* report fails to acknowledge,[23] and that such alternatives receive less attention in debates than the optimistic growth-driven framing of a green economy.[24]

There are, furthermore, diverging views about the role of markets in a green economy. Market-based approaches see the current economic system as more or less fit to deal with environmental harm and promote a commodification of nature and payment for ecosystem services.[25] However, the eco-economic management that can underpin such approaches has been criticised, because it may cement inequalities in favour of global elites and because it overlooks social differences on the ground.[26] Furthermore, there are concerns about the market's ability to deal with environmental issues in an effective and equitable manner and about accurate pricing of externalities.[27] Kosoy et al go so far as to contend that sustainability cannot be achieved under the premises of neo-classical economics.[28] Systemic change is consequently argued for.[29] These fundamentally different perceptions of the potential of markets in a green economy contribute to the broad range of green economy notions, from radical change to a mere adaptation of the economic system.

It is crucial not to forget the political implications when immersed in debates to elucidate meaning. UNEP's Green Economy initiative sets out to transition the economy to address poverty and ensure sustainability in the 21st century.[30] A green economy, Brand contends, 'seems to promise an attractive orientation out of the crisis of neoliberalism that became manifest in 2008'.[31] From a political economy perspective the green economy and green growth have more critical political implications. Wanner argues that, by promising to decouple economic growth from environmental impacts, the green economy sustains neoliberal capitalism.[32] The green economy, from this perspective, is not so much about change but, in Gramsci's terms, about reinforcing hegemony through a 'passive revolution'.[33]

Analysing tourism and the green economy

The diverging perspectives presented above demonstrate that the green economy can be rhetorically and materially constructed on the basis of greatly diverging premises.[34] This opens up two avenues of inquiry for this paper. First, the paper explores how UNEP and the United Nations World Tourism Organization (UNWTO), as well as a number of industry representatives, comprehend the green economy, their shared understandings and assumptions, and the norms and values they promote.[35] This also necessitates being open to alternative understandings and normative visions of a green economy that are marginalised.[36] Second, the paper assesses whether the pathway towards a green economy that industry representatives envisage can be implemented. Exploring rhetoric and implementation is crucial, as UNEP's *Towards a Green Economy* has been accused of resembling a 'science fiction novel at times',[37] while tourism research, as the paper shows below, suggests that at least some of the premises on which it builds are unsound.

The paper draws on two documentary sources: *Towards a Green Economy* and *Green Growth and Travelism*, as well as peer-reviewed research to examine these questions in a thematic analysis.[38] *Green Growth and Travelism* contains 46 letters from leaders in the tourism industry, who represent a range of backgrounds and include representatives from UN organisations, industry associations, airlines, destinations and universities. The two documents are the most comprehensive sources of how UNEP, the UNWTO and a broad range of leading industry actors understand and envisage tourism's role in a green economy. They convey political agendas and practice recommendations and as such represent discursive practices that can shape governance by promoting certain rules and norms of climate change mitigation.[39] *Towards a Green Economy* makes the case for investments in greening tourism. *Green Growth and Travelism* states that it 'explores why the industry is misperceived and how it can take its rightful leadership place in the transformation to the new green economy'.[40] While attention to their production might reveal more insights into power relations and the marginalisation of alternative approaches to climate change,[41] this analysis focuses on the final documents to better understand the messages that they convey.

The documents were analysed thematically, which entailed inductive coding and recoding as well as developing and reviewing themes.[42] Attention was paid to how frequently individual codes and themes appeared in the documents, keeping in mind that the book that compiles 46 letters from leaders was edited to 'minimise repetition'.[43] The key themes are presented in Figure 1 and

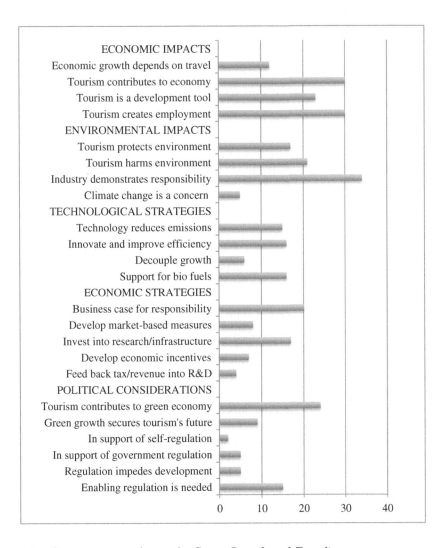

Figure 1.   Green economy themes in *Green Growth and Travelism*.

referred to throughout the text. There are some limitations to this approach. This article does not offer sufficient space to do justice to every single notion and the wide-ranging solutions proposed in the 46 letters; it attempts, however, to convey the key arguments. It also needs to be noted that, while the authors who contributed to *Green Growth and Travelism* represent many areas of tourism, they are a selection of actors and more extensive research would be necessary to provide a broader representation of tourism stakeholders.

In what follows I first outline how tourism's role in the green economy is perceived. The paper then explores the case for tourism as a development tool, and related criticism. This leads the paper to expand on the implications for tourism in times of its increasing climate change contributions, before it examines how green growth is to be achieved. Again, industry claims are linked to research that casts doubt on the industry's green growth agenda. The paper concludes by arguing that green growth stands on shaky foundations and represents a political discourse that promotes further growth and marginalises radical change, despite convincing environmental concerns.

Tourism in the green economy

With the integration of tourism into *Towards a Green Economy*, UNEP encourages the sector to shift to sustainable tourism, by which it refers to 'tourism activities that can be maintained, or sustained, indefinitely in their social, economic, cultural and environmental contexts'.[44] *Green Growth and Travelism* (and Figure 1) leave little doubt that UN organisations, industry associations and businesses have, at least rhetorically, absorbed tourism's vital role in the process towards a green economy. This confidence is driven by an underlying conviction that tourism can contribute to development, poverty alleviation and employment, while reducing environmental degradation.[45] Several authors acknowledge negative environmental and social impacts and the need to take into account environmental limits.[46] The overall sense, however, is that, if done right and with appropriate investments and policies, tourism can contribute to growth that is both just and inclusive.[47]

Apart from creating benefits for environment and society, sustainable tourism in a green economy is associated with business opportunities. Gloria Guevara, Secretary of Tourism in Mexico, argues that growing sustainably is cheaper for destinations than adapting to impacts of climate change.[48] Additionally, some authors contend that sustainable tourism offers marketing opportunities for developing countries,[49] although early engagement by UNWTO and the European Travel Commission with the topic specifies that efforts should be driven by environmental and social concerns rather than marketing benefits.[50] Apart from financial considerations, there are also important legitimacy implications. This clearly comes out in Angela Gittens' letter on airport expansion. The Director General of Airports Council International argues that growth relies on air transport and this requires social legitimacy that can be brought about by past and future environmental stewardship.[51] Promoting a green economy is therefore also about preparing the tourism business for the future.

This ability to grow while greening the business defines growth-centred understandings of the green economy, such as a 'thin green economy' and 'green growth.'[52] It is also evident in *Green Growth and Travelism*, which departs from UNEP's terminology of a 'green economy' and frames industry's efforts around 'green growth'. This focus is not uncommon. Within the context of the green economy it can frequently be observed that development is substituted by the term 'growth'.[53] This framing indicates that 'greening' is essential, but that this should not impede growth; it reflects an understanding of the green economy as a discursive project to protect capitalism.[54] The editors of *Green Growth and Travelism*, however, caution criticism. They argue that aviation should not receive blame outright for its climate change contribution, but that efforts to reduce emissions and tourism's crucial benefits for development need to be appreciated.[55] It is to an analysis of tourism as a development tool that this paper turns next.

Tourism as a development tool in a green economy

Global development and equity are defining components of a green economy, both in theory and in practice. UNEP promotes tourism as a sector that can contribute to the agenda, because it can be designed to benefit destination economies and their environments.[56] Developmental benefits also feature prominently in *Green Growth and Travelism*.[57] Both documents highlight how tourism operators create such benefits by protecting natural resources in destinations and by creating the following employment and economic opportunities.

The green economy discourse embraces employment opportunities and the associated redistribution of wealth. *Towards a Green Economy* and most letters in *Green Growth and Travelism* highlight the fact that almost 10% of global jobs are tourism-related.[58] Through trickle-down effects, incomes from tourism activities are considered to benefit the wider economy.[59] In line with these positive outcomes many letters affirm that tourism can lead to empowerment and lift people out of poverty.[60]

There is also frequent reference to tourism's contribution to development; it adds to gross domestic product and foreign exchange earnings. Several letters note that one in three developing countries and half of the least developed countries have tourism as their main source of foreign exchange.[61] Twenty-three out of 49 of the least developed countries are reported to have tourism among the three largest revenue earners, while for seven it is the largest source of revenue.[62] Taleb Rifai, the Secretary-General of UNWTO, considers tourism to be one of the few opportunities for the least developed countries to partake in the global economy,[63] while Lucian Cernat (European Commission) argues that tourism provides these countries with the resources to diversify their economic structure.[64] By its very nature, it appears, tourism contributes to creating the more equal society that a green economy aims for.

Finally, the documents highlight tourism's contribution to intercultural understanding as a result of interactions between people from different backgrounds. Marthinus Van Schalkwyk, Minister of Tourism in South Africa, points out that tourism promotes peace and harmony and is a suitable economic activity for rebuilding after wars and conflicts.[65] Tony Tyler, Director General and CEO of the International Air Transport Association (IATA), argues that today's global economy has become reliant on aviation and its ability to connect the world to foster personal, economic and cultural exchange, as well as trade.[66] Travel, in short, has become the basis of our socioeconomic lives.

Both *Towards a Green Economy* and *Green Growth and Travelism* present the tourism industry as one that, thanks to its positive economic outcomes in destinations, is suited to transform not only the tourism economy, but also associated economies. While there are some notes of caution, the overall narrative is that tourism's role in development and alleviating poverty is a positive one.[67] Sustainable and responsible growth is wanted. Restrictions on tourism flows, on the other hand, as several letters point out, would curtail the positive benefits that the industry can bring about.[68]

Challenging tourism's potential for development

However, critical research suggests that tourism's contribution to development and poverty alleviation is far from clear. While there is no doubt that positive examples exist, Chok et al. consider current evidence to be insufficient to draw conclusive correlations.[69] Apart from uneasiness about environmental degradation and social impacts as a result of tourism activities,[70] researchers have raised concerns relating to the equity aspects of employment and economic opportunities that challenge the green economy initiative that UNEP, UNWTO and industry actors promote.

While this critical body of research acknowledges that tourism provides employment opportunities that might not otherwise exist,[71] the concern is whether these positions contribute to equity and poverty alleviation. Labour standards in tourism, Chok et al point out, can be precarious and exploitative.[72] Tourism employment is characterised by 'low wages, over-dependency on tips, long working hours, stress, lack of secure contracts, poor training and almost no promotion opportunity'.[73] Overcoming poverty under such conditions seems improbable, as they do not permit any accumulation of wealth and offer no safety net.[74] Scheyvens argues that this situation may be aggravated by the weak standing that destination communities have in comparison to external entities that shape development.[75] Furthermore, inequalities within destinations must not be overlooked, as these can lead to elite capture of decision-making processes and the benefits from tourism.[76] Relatively richer members of societies with the means to invest and operate tourism services can be expected to benefit more than poorer members.[77] Geographic differences in tourism at the national scale may also complicate equitable development.[78] In the context of overcoming inequalities, leakage of tourism spending is a crucial factor. While, for some communities, tourism income despite leakage exceeds income from other activities,[79] researchers estimate that, on average, 40% to 50% of gross tourism earnings are not retained in destinations.[80] These concerns around employment and the redistribution of wealth cast doubt on the potential for a green economy. There is, rather, a risk that tourism growth aggravates existing local, national and global inequalities.

For now the industry's contribution to development remains controversial, given that sufficient empirical evidence to prove either case is lacking.[81] To resolve this issue, Gössling et al argue for a more nuanced analysis of the potential of tourism in development and poverty alleviation, which goes beyond GDP as a measure of success and takes into account socioeconomic and environmental factors.[82] They argue that higher arrival numbers do not necessarily result in better livelihood outcomes.[83] This correlation, or lack thereof, is crucial for the feasibility of a green economy and Chok et al raise concerns about promoting tourism development based on the assumption that tourism is a solution for sustainability issues.[84] Developing tourism might not only aggravate negative impacts on the ground; it will also increase emissions and further increase tourism's climate change impact.

Implications for a green economy

Awareness of these contradictions is crucial, because tourism's role as a development tool is employed in the climate change discourse and consequently fosters rules and norms of climate change mitigation that are based on foundations that some researchers question. UNWTO, for example, refers to tourism's contribution to poverty alleviation and development in order to argue for special treatment of aviation to least developed countries in the climate change context.[85] Likewise the editors of *Green Growth and Travelism* caution that, because tourism benefits developing countries, aviation should not be demonised.[86] These arguments promote a normative context that the tourism industry should not be as harshly criticised for the side-effect of climate change when offering development benefits. This framing differs notably from the more urgent case that Hall et al make, namely that mitigation will be difficult and climate change 'effects will potentially cancel the supposed benefits of the pursuit of international tourism in many cases'.[87] If green growth obfuscates these contradictions, we need to raises the crucial question of how those promoting green growth intend to achieve emissions reductions.

Green growth is low-carbon growth

There is no disagreement that, for tourism to grow, it needs to reduce its emissions. *Towards a Green Economy* and *Green Growth and Travelism* present initiatives to demonstrate that emissions reductions are feasible and tourism businesses are on a successful path towards change. UNEP estimates that, under the green investment scenario, CO_2 emission reductions in the order of 52% compared to the business-as-usual scenario are possible.[88] In *Green Growth and Travelism* several letters refer to the targets of IATA and the World Travel and Tourism Council (WTTC) to halve 2005 CO_2 emissions levels by 2050 and 2035, respectively.[89] Fundamental to these emissions reductions are technological innovation, biofuels and market-based measures that find broad support from industry stakeholders, as many letters in *Green Growth and Travelism* evidence.

The framing of technological innovation is an optimistic one that expects future emissions reductions based on past experience.[90] While many letters call for technological progress, they remain notably unclear about the precise development that they expect. Overall there appears to be a sense that progress is essential in order to transition to a green economy, as it decouples emissions from growth.[91] A modern fleet of aircraft, for example, is seen as a crucial component of reducing emissions.[92] Improving efficiency is seen to make good business sense, as it reduces fuel costs.[93] While overall much hope lies in technology, some authors are more cautious in their evaluation. Giovanni Bisignani points out emissions have been decreasing, noting, however, that emissions reductions of 3% annually have been surpassed by the growth in air traffic of 5%.[94] This statement encapsulates the challenge for many tourism businesses, to which this paper returns below: a positive framing of technological potential that implies manageability of emissions, but one that marginalises the environmental implications of increasing tourism flows.

Biofuels are promoted as a critical technological initiative to drive the green transition. Several letters highlight the progress achieved so far in terms of certification and commercial testing.[95] While issues concerning the environmental impacts and land-use pressures of first-generation biofuels are noted, the focus of the green growth discourse is on second-generation biofuels with their improved sustainability performance.[96] Commercialisation and scaling up are seen as areas of current shortcoming; nonetheless the overall framing is optimistic.[97] Tony Tyler, Director General and CEO of IATA, expects a reduction of the carbon footprint in the order of 80%, which he argues would contribute substantially to securing the legitimacy for aviation to grow.[98] Rhetorically biofuels are emerging as a promising avenue for emissions reductions.

In order to secure the transition towards a green economy, many letters argue in line with UNEP's *Towards a Green Economy* that tourism businesses require investment in research, education and infrastructure.[99] UNWTO Secretary-General Taleb Rifai makes a strong case for investment in greening tourism activities to harness growth, employment and development opportunities and to reduce emissions. He argues that, in particular, investment in small and medium enterprises and in developing countries is essential, as these businesses and countries lack the resources for greening tourism activities. By pursuing investment in green strategies, Rifai expects that the tourism industry will be able to expand and prosper in the future.[100] However, while investment in greening tourism is bound to speed up change, it also alters responsibilities for addressing environmental issues: it assigns governments the role of supporting businesses in their private efforts to reduce emissions.

In addition to support for technological innovation, the transition towards a green economy is often associated with market-based measures, although the shape this is to take remains a contentious issue. Many letters criticise the current unstructured taxation of aviation, which is seen to channel substantial amounts of funding away from climate change mitigation, thereby depriving private actors of funding to invest in mitigation.[101] Some letters even argue that industry is already contributing more than necessary in tax to compensate for aviation's emissions.[102] Several authors in *Green Growth and Travelism* therefore recommend that funds gained from auctioning emissions certificates should flow back into research and the development of technologies.[103]

The recent shifts from taxation to cap and trade mechanisms occupy many authors in *Green Growth and Travelism* and reveal conflicting standpoints. Marthinus Van Schalkwyk, Minister of Tourism in South Africa, argues for binding regulation with non-compliance measures and a stringent carbon cap, building on an understanding that the industry's future success requires speeding up the process of decarbonising aviation.[104] James Hogan, the president and

CEO of Etihad Airways, on the other hand, considers market-based measures as a last resort, to be used only if alternative initiatives fail to meet emissions targets.[105] These conflicting positions on market-based measures are challenges that the International Civil Aviation Organization (ICAO) now faces in developing a global measure to address emissions. While the outcome of ICAO's global market-based measure remains open, this recent turn towards market-based solutions is in line with more widely pursued governance approaches to climate change mitigation that reinforce neoliberal capitalism.

The overall outlook for emissions reductions in *Green Growth and Travelism* and *Towards a Green Economy* is optimistic and presents the industry as one with great potential to reduce its carbon footprint. Technological innovation, and to a large part market-based measures, are presented as crucial pathways towards climate change mitigation. A rather different perspective emerges from recent research activities.

Challenging low-carbon growth potential

Recent research on the potential for emissions reduction in tourism activities is less optimistic than industry's framing. Here, the tourism industry is already seen to be a major contributor to climate change, emitting 5% of global CO_2 emissions in 2005 and contributing up to 12.5% to radiative forcing, a cumulative measure of climate change contribution which includes non-CO_2 effects.[106] Rather than seeing the potential for substantial emissions reductions, researchers have voiced concerns that technological limitations in times of significant growth of tourism flows and increasing travel distances make absolute emissions reductions unlikely.[107]

Several researchers view the technological innovations that industry foresees in support of green growth more critically. Although they do not deny that efficiency improvements have significantly reduced relative emissions, they point out that the projections for the future are overly optimistic and deconstruct industry's claims.[108] In terms of absolute emissions reductions it is doubtful how extrapolated efficiency improvements in the range of 1%–2% annually will reduce emissions from a 3.3% annual increase in tourism demand.[109] Lee et al point out that, in order to achieve absolute emissions reductions, radical technological innovations are essential, as many opportunities around airframe aerodynamics and materials have already been realised; commercial implementation of new technologies, however, is a time-consuming process.[110] The promise of biofuels as an alternative fuel source is also viewed with caution. Timilsina and Shrestha point out substantial sustainability issues associated with competing land use for agriculture and the ecological effects of converting land. While industry actors urge rapid commercialisation and scaling up, their evaluation expects this to be unlikely.[111] The technological potentials promoted by UNEP and industry actors appear optimistic in light of these research findings.

The slow progress expected on the technological front has led researchers to question whether the industry can reduce increasing emissions from quickly growing tourism flows. Even a UNWTO-UNEP-WMO commissioned report expects that, between 2005 and 2035, CO_2 emissions could increase by 161%, based on a 4% annual growth in international arrivals.[112] If business continues as usual, Scott et al's model shows that around 2050 or 2060 emissions from tourism activities alone could exceed the global sustainable budget and thereby challenge global efforts to avoid dangerous climate change.[113] Therefore, if the tourism industry wants to be in line with other sectors to avoid dangerous climate change, a transformation is required. This transformation would need to combine modal shift, shorter travel distances, a quick introduction of low-carbon technologies and demand management.[114] Dubois et al go so far as to suggest that, eventually, there need to be considerations about 'who can travel, for how long, using which transport mode, why, and how comfortably'.[115] These questions raise a highly controversial normative decision on the right to travel that in a carbon-constrained world will need to be made, unless emissions can be cut radically. In particular, the questions draw attention to the current geographic imbalance in travel and the growing demand of an increasingly affluent middle class in developing countries that may demand its right to travel. Given current trajectories of tourism flows in developed and developing countries, industry's green growth projections appear highly unrealistic.

These radically different perspectives on the potential for emissions reductions are not only ecologically disconcerting; they entail political implications that are vastly different. Based on the research cited above, the green growth-driven notion of the green economy that industry promotes appears an unlikely solution to reducing emissions from tourism sufficiently to avoid dangerous climate change. A more radical green economy understanding that reconsiders growth, like the 'thick green economy' or 'green revolution', seems necessary, unless industry can demonstrate its forecast potential against the doubts that research raises.[116] If industry's envisaged change cannot be achieved, green growth becomes a risky discourse that hides conflicts between economic and environmental sustainability in favour of sustaining economic growth. It would present a strategy that Wanner sees as a 'passive revolution', through which neoliberal capitalism adjusts in response to environmental crisis, but one that endangers the global climate.[117]

Conclusion

By exploring the presentation of the green economy in *Towards a Green Economy* and *Green Growth and Travelism* this paper has highlighted fundamental conflicts in their framing of green economy and green growth and some researchers' expectations pertaining to the transition towards a green economy. The paper has shown how industry understands the green economy as one that creates economic, developmental and environmental benefits on the basis of sustainable tourism, for which it encourages a supportive environment. This notion is rhetorically most similar to Ehresman and Okereke's 'thin green economy' in that it fosters poverty alleviation, equality and environmental improvement. And it reflects Death's 'green transformation', which does not challenge the current economic model, but calls for investments and incentives for a socially just and equitable economy.[118] However, by drawing on tourism research this paper has also pointed out that, materially, these framings may be overly optimistic. Tourism research identifies clear material shortcomings relating to environmental and developmental issues that need to be resolved, especially given the forecast growth.[119] This is all the more important as the growth in tourism and its associated contribution to climate change may cancel out some of the benefits in developing countries that a green economy promises.[120] Crucially, and beyond these material shortcomings, this paper has emphasised the political implications of the green economy discourse for climate change governance in the tourism sector.

The green economy framing sets the tourism industry on a controversial path towards future growth by promoting rules and norms of climate change mitigation that reflect an adjustment of neoliberal capitalism to the environmental crisis that has been criticised more broadly in terms of the green economy.[121] The framing does not question the existing business model that causes environmental harm. Rather, it provides the legitimacy for tourism to grow by presenting tourism as a strategy for poverty alleviation, employment and income generation for communities in developing countries, and by highlighting the potential to reduce the negative impacts of tourism activity. However, if the tourism industry fails to contribute to development and emissions increase, a direction that the researchers referred to in this paper expect to be likely as a result of growing demand,[122] the impacts on the environment and societies in destinations could be detrimental.

For now there is no certainty as to whether the tourism industry will be able to deliver, making it all the more important to consider who might gain from the rules and norms that the green economy discourse fosters. Identifying who benefits is, however, a tricky task. UNEP's initiative is one that provides guidance, but no clear agenda for implementation. Industry is yet to prove that it can deliver the benefits it foresees. Therefore, in order to understand who benefits from a green economy, it is worthwhile reversing the question to ask about the implications if tourism flows are transformed, ie regulated in order to achieve the substantial emissions reductions that avoiding dangerous climate change requires. The European Commission suggests that global emissions in 2050 need to be reduced by 90% against 1990 emissions levels.[123] This would require de-growing tourism activities, an approach that is implied in the proposals put forward by several researchers on how to transform the sector.[124] Yet only a few researchers, like Hall in his publication *Degrowing Tourism*, explicitly employ the term.[125] The consequences of de-growing tourism activities would be far-reaching. It would challenge the strong focus on growth that dominates many tourism businesses and organisations like UNWTO and WTTC. Crucially it would restrict the opportunities of many tourism businesses, in particular those reliant on aviation, and would leave many tourism destinations deprived of existing business opportunities. De-growth would mean severely altering consumer travel patterns and would require sacrificing material conveniences, like flying, to live within ecological limits.[126] It is a development that will probably not be in the interest of tourism operators, destinations and consumers; but it is one that these researchers, based on the best scientific knowledge, deem essential. The green growth notion promoted by tourism stakeholders makes such radical changes redundant by concealing contradictions between the potential for emissions reduction and growth. It is a 'passive revolution' that benefits business actors and their representatives when addressing the challenge of climate change that their business activities have contributed to.

These contradictions and political implications are particularly disconcerting, as they come at a time when the post-Kyoto policy process and ICAO's development of global market-based measures are ongoing. The green growth discourse informs these policy processes and may shift them (and other emerging debates) in a way that could avert radical change and maintain a stable business environment for a growing tourism industry – albeit by promoting growth based on the contentious foundations of development benefits and environmental potential.

Funding

This work was supported by the Sustainable Consumption Institute.

Notes

1. UNEP, *Towards a Green Economy*, 420.
2. Lipman et al., *Green Growth and Travelism*. The term 'travelism' is short for travel and tourism.
3. Ehresman and Okereke, "Environmental Justice"; and Death, "The Green Economy in South Africa."
4. UNEP, *Towards a Green Economy*, 420.
5. Gössling et al., "The Challenges of Tourism as a Development Strategy"; Hall et al., "The Primacy of Climate Change"; Peeters and Eijgelaar, "Tourism's Climate Mitigation Dilemma"; and Scott et al., "Can Tourism Deliver?"
6. Bailey and Caprotti, "The Green Economy"; Pearce et al., *Blueprint for a Green Economy*; Death, "The Green Economy in South Africa"; and Brown et al., "Green Growth or Ecological Commodification."
7. UNEP, *Green Economy*, 5.
8. UNEP, *Towards a Green Economy*, 16.
9. Dryzek, *The Politics of the Earth*, 9.
10. Bowen and Frankhauser, "The Green Growth Narrative."
11. UNEP, *Towards a Green Economy*.
12. Cook and Smith, "Introduction."
13. Tienhaara, "Varieties of Green Capitalism"; and Brand, "Green Economy."
14. Ehresman and Okereke, "Environmental Justice," 17.
15. Death, "The Green Economy in South Africa," 2.
16. Bartelmus, "The Future we Want"; Borel-Saladin and Turok, "The Green Economy"; Cook and Smith, "Introduction"; Tienhaara, "Varieties of Green Capitalism"; and UNEP, *Towards a Green Economy*.
17. Kosoy et al., "Pillars for a Flourishing Earth."
18. This notion draws on Meadows and Meadows, *The Limits to Growth*.
19. Kosoy et al., "Pillars for a Flourishing Earth"; and Kallis, "In Defence of Degrowth."
20. Kosoy et al., "Pillars for a Flourishing Earth."
21. Harcourt, "Editorial."
22. Brand, "Green Economy."
23. Brockington, "A Radically Conservative Vision?"
24. Brand, "Green Economy"; and Bowen and Frankhauser, "The Green Growth Narrative."
25. Pearce et al., *Blueprint for a Green Economy*; and Barbier and Markandya, *A New Blueprint for a Green Economy*.

26. Harcourt, "Editorial."
27. Ibid; and Kosoy et al., "Pillars for a Flourishing Earth."
28. Kosoy et al., "Pillars for a Flourishing Earth."
29. Jessop, "Economic and Ecological Crises"; Harcourt, "Editorial"; and Kosoy et al., "Pillars for a Flourishing Earth."
30. UNEP, *Towards a Green Economy*, 7.
31. Brand, "Green Economy," 28.
32. Wanner, "The New 'Passive Revolution'."
33. Gramsci, *Selections from the Prison Notebooks*.
34. Bailey and Caprotti, "The Green Economy."
35. Ibid; and Dryzek, *The Politics of the Earth*.
36. Cook and Smith, "Introduction"; and Bailey and Caprotti, "The Green Economy."
37. Brockington, "A Radically Conservative Vision?," 410.
38. Braun and Clark, "Using Thematic Analysis."
39. Bulkeley and Newell, *Governing Climate Change*; and Dryzek, *The Politics of the Earth*.
40. Lipman et al., *Green Growth and Travelism*, back cover.
41. Atkinson and Coffey, "Analysing Documentary Realities"; and Dryzek, *The Politics of the Earth*.
42. Braun and Clark, "Using Thematic Analysis."
43. Lipman et al., "Introduction and Overview," 2.
44. UNEP, *Towards a Green Economy*, 420.
45. Lipman et al., *Green Growth and Travelism*.
46. Scowsill, "Travel and Tourism leading the Way"; Francis, "Greening the Tourism Sector"; de Villiers, "Responsible Tourism"; van Schalkwyk, "Breaking out of Silos"; and Cernat, "Tourism Performance Benchmarking."
47. Rifai, "One Billion Tourists."
48. Guevara, "Tourism, Green Growth and Sustainable Development."
49. Francis, "Greening the Tourism Sector."
50. UNWTO, European Travel Commission and VisitSweden, *Travel and Tourism in the Green Economy Symposium Conclusions*.
51. Gittens, "Aviation and Green Growth."
52. Ehresman and Okereke, "Environmental Justice," 17; and Death, "The Green Economy in South Africa," 2.
53. Bailey and Caprotti, "The Green Economy."
54. Wanner, "The New 'Passive Revolution'."
55. Lipman et al., "Towards Transformation."
56. UNEP, *Towards a Green Economy*.
57. Lipman et al., *Green Growth and Travelism*.
58. Bezbaruah, "Everybody's Action or Nobody's Action?"; de Villiers, "Responsible Tourism"; and van Schalkwyk, "Breaking out of Silos."
59. Bezbaruah, "Everybody's Action or Nobody's Action?"; and Branson, "Why Aviation can Respond Positively to Climate Change."
60. Bezbaruah, "Everybody's Action or Nobody's Action?"; Rifai, "One Billion Tourists"; and Tyler, "A Green Economy."
61. Francis, "Greening the Tourism Sector"; and Lefebvre, "The Green Economy."
62. Panitchpakdi, "Sustainable Tourism Development."
63. Rifai, "One Billion Tourists."
64. Cernat, "Tourism Performance Benchmarking."
65. van Schalkwyk, "Breaking out of Silos."
66. Tyler, "A Green Economy."
67. Branson, "Why Aviation can Respond Positively to Climate Change"; de Villiers, "Responsible Tourism"; Francis, "Greening the Tourism Sector"; Rifai, "One Billion Tourists"; and van Schalkwyk, "Breaking out of Silos."
68. Lyle, "Aviation's Role in Green Growth."
69. Chok et al., "Tourism as a Tool for Poverty Alleviation."
70. Newsome et al., *Natural Area Tourism*; and Duffy, *Nature Crime*.
71. Sharpley, *Tourism Development and the Environment*.
72. Chok et al., "Tourism as a Tool for Poverty Alleviation."
73. Beddoe, cited in Chok et al., "Tourism as a Tool for Poverty Alleviation," 157.
74. Chok et al., "Tourism as a Tool for Poverty Alleviation."
75. Scheyvens, "Ecotourism and the Empowerment of Local Communities."
76. Schilcher, "Growth versus Equity"; and Sharpley, *Tourism Development and the Environment*.
77. Blake, "Tourism and Income Distribution in East Africa."
78. Oppermann, "Tourism Space in Developing Countries."
79. Sandbrook, "Putting Leakage in its Place."

80. UNCTAD, *Sustainable Tourism*. Leakage is also noted as a problem in Bezbaruah, "Everybody's Action or Nobody's Action?"; Cernat, "Tourism Performance Benchmarking"; and Dodds, "Traveling to Tomorrow."
81. Chok et al., "Tourism as a Tool for Poverty Alleviation"; Gössling et al., "The Challenges of Tourism as a Development Strategy"; and Saarinen et al., "Tourism and Millennium Development Goals."
82. Gössling et al., "The Challenges of Tourism as a Development Strategy."
83. Ibid.
84. Chok et al., "Tourism as a Tool for Poverty Alleviation."
85. UNWTO, *From Davos to Copenhagen and Beyond*.
86. Lipman et al., "Towards Transformation."
87. Hall et al., "The Primacy of Climate Change," 118.
88. UNEP, *Towards a Green Economy*.
89. Bisignani, "The Mystery of Governments' Aviation Policy"; Gittens, "Aviation and Green Growth"; Hogan, "Green Growth and Travelism"; Scowsill, "Travel and Tourism leading the Way"; and Tyler, "A Green Economy."
90. Benjamin, "Aviation's and ICAO's Climate Change Response"; and Tyler, "A Green Economy."
91. Hogan, "Green Growth and Travelism"; and Bisignani, "The Mystery of Governments' Aviation Policy."
92. Gittens, "Aviation and Green Growth."
93. Bisignani, "The Mystery of Governments' Aviation Policy"; Branson, "Why Aviation can Respond Positively to Climate Change"; Enders, "One Born Every Minute"; and van Schalkwyk, "Breaking out of Silos."
94. Bisignani, "The Mystery of Governments' Aviation Policy."
95. Benjamin, "Aviation's and ICAO's Climate Change Response"; Bisignani, "The Mystery of Governments' Aviation Policy"; Gittens, "Aviation and Green Growth"; and Tyler, "A Green Economy."
96. Bisignani, "The Mystery of Governments' Aviation Policy."
97. Benjamin, "Aviation's and ICAO's Climate Change Response"; Branson, "Why Aviation can Respond Positively to Climate Change"; and Tyler, "A Green Economy."
98. Tyler, "A Green Economy."
99. Branson, "Why Aviation can Respond Positively to Climate Change"; Enders, "One Born Every Minute"; and Hogan, "Green Growth and Travelism."
100. Rifai, "One Billion Tourists."
101. Hogan, "Green Growth and Travelism"; Ambrose, "A Fairer Environmental Balance"; and Bisignani, "The Mystery of Government's Aviation Policy." See also Lyle, "Aviation's Role in Green Growth."
102. Bisignani, "The Mystery of Governments' Aviation Policy"; Hogan, "Green Growth and Travelism"; and Tyler, "A Green Economy."
103. Ambrose, "A Fairer Environmental Balance."
104. van Schalkwyk, "Breaking out of Silos."
105. Hogan, "Green Growth and Travelism."
106. Scott et al., "Can Tourism Deliver?"; and Hall et al., "The Primacy of Climate Change."
107. Hall et al., "The Primacy of Climate Change"; Hall, "Framing Behavioural Approaches"; and Peeters and Dubois, "Tourism Travel under Climate Change Mitigation Constraints."
108. Gössling et al., "The Future of Tourism"; Gössling and Peeters, "'It does not Harm the Environment!'"; and Hall et al., "The Primacy of Climate Change."
109. IPCC, *IPCC 4th Assessment Report*; Lee, "Can we Accelerate the Improvement of Energy Efficiency?"; and UNWTO, *UNWTO Tourism Highlights*.
110. Lee et al., "Aviation and Global Climate Change."
111. Timilsina and Shrestha, "How much Hope?"
112. UNWTO et al., *Climate Change and Tourism*.
113. Scott et al., "Can Tourism Deliver?"
114. Dubois et al., "The Future Tourism Mobility"; and Peeters and Dubois, "Tourism Travel under Climate Change Mitigation Constraints."
115. Dubois et al., "The Future Tourism Mobility," 1041.
116. Ehresman and Okereke, "Environmental Justice", 17; and Death, "The Green Economy in South Africa", 2.
117. Wanner, "The New 'Passive Revolution'." Wanner draws on Gramsci's understanding of a 'passive revolution' contained in *Selections from the Prison Notebooks*.
118. Ehresman and Okereke, "Environmental Justice"; and Death, "The Green Economy in South Africa."
119. In particular Hall et al., "The Primacy of Climate Change."
120. Ibid.
121. Wanner, "The New 'Passive Revolution'."
122. Scott et al., "Can Tourism Deliver?"
123. European Commission, "Communication from the Commission to the European Parliament."
124. Dubois et al., "The Future Tourism Mobility"; and Peeters and Dubois, "Tourism Travel under Climate Change Mitigation Constraints."

125. For an exception, see Hall, "Degrowing Tourism."
126. This point draws on Wapner's conception of sacrifice. Wapner, "Sacrifice."

Bibliography

Ambrose, M. "A Fairer Environmental Balance – Aviation's Responsible Role." In *Green Growth and Travelism: Letters from Leaders*, edited by G. Lipman, T. Delacy, S. Vorster, R. Hawkins and M. Jiang, 230–235. Oxford: Goodfellow Publishers, 2012.

Atkinson, P., and A. Coffey. "Analysing Documentary Realities." In *Qualitative Research: Theory, Method and Practice*, edited by D. Silverman, 56–75. London: Sage, 2004.

Bailey, I., and F. Caprotti. "The Green Economy: Functional Domains and Theoretical Directions of Enquiry." *Environment and Planning A* 46, no. 8 (2014): 1797–1813.

Barbier, E., and A. Markandya. *A New Blueprint for a Green Economy*. Abingdon: Routledge, 2013.

Bartelmus, P. "The Future we Want: Green Growth or Sustainable Development?" *Environmental Development* 7 (2013): 165–170.

Benjamin, R. "Aviation's and ICAO's Climate Change Response." In *Green Growth and Travelism: Letters from Leaders*, edited by G. Lipman, T. Delacy, S. Vorster, R. Hawkins and M. Jiang, 242–245. Oxford: Goodfellow Publishers, 2012.

Bezbaruah, M. P. "Everybody's Action or Nobody's Action? Rio+20, Tourism and India." In *Green Growth and Travelism: Letters from Leaders*, edited by G. Lipman, T. Delacy, S. Vorster, R. Hawkins and M. Jiang, 246–253. Oxford: Goodfellow Publishers, 2012.

Bisignani, G. "The Mystery of Governments' Aviation Policy." In *Green Growth and Travelism: Letters from Leaders*, edited by G. Lipman, T. Delacy, S. Vorster, R. Hawkins and M. Jiang, 254–260. Oxford: Goodfellow Publishers, 2012.

Blake, A. "Tourism and Income Distribution in East Africa." *International Journal of Tourism Research* 10, no. 6 (2008): 511–524.

Borel-Saladin, J. M., and I. N. Turok. "The Green Economy: Incremental Change or Transformation?" *Environmental Policy and Governance* 23, no. 4 (2013): 209–220.

Bowen, A., and S. Frankhauser. "The Green Growth Narrative: Paradigm Shift or Just Spin?" *Global Environmental Change* 21, no. 4 (2011): 1157–1159.

Brand, U. "Green Economy – The Next Oxymoron? No Lessons Learned from Failures of Implementing Sustainable Development". GAIA: Ecological Perspectives for." *Science and Society* 21, no. 1 (2012): 28–32.

Branson. R. "Why Aviation can Respond Positively to Climate Change." In *Green Growth and Travelism: Letters from Leaders*, edited by G. Lipman, T. Delacy, S. Vorster, R. Hawkins and M. Jiang, 261–265. Oxford: Goodfellow Publishers, 2012.

Braun, V., and V. Clark. "Using Thematic Analysis in Psychology." *Qualitative Research in Psychology* 3, no. 2 (2006): 77–101.

Brockington, D. "A Radically Conservative Vision? The Challenge of UNEP's Towards a Green Economy." *Development and Change* 43, no. 1 (2012): 409–422.

Brown, E., J. Cloke, D. Gent, P. H. Johnson, and C. Hill. "Green Growth or Ecological Commodification: Debating the Green Economy in the Global South." *Geografiska Annaler: Series B, Human Geography* 96, no. 3 (2014): 245–259.

Bulkeley, H., and P. Newell. *Governing Climate Change*. Abingdon: Routledge, 2010.

Cernat, L. "Tourism Performance Benchmarking for Sustainability Development Policy." In *Green Growth and Travelism: Letters from Leaders*, edited by G. Lipman, T. Delacy, S. Vorster, R. Hawkins and M. Jiang, 271–277. Oxford: Goodfellow Publishers, 2012.

Chok, S., J. Macbeth, and C. Warren. "Tourism as a Tool for Poverty Alleviation: A Critical Analysis of 'Pro-poor Tourism' and Implications for Sustainability." *Current Issues in Tourism* 10, no. 2 (2007): 144–165.

Cook, S., and K. Smith. "Introduction: Green Economy and Sustainable Development: Bringing back the 'Social'." *Development* 55, no. 1 (2012): 5–9.

de Villiers, D. "Responsible Tourism, Local Communities and Ethics." In *Green Growth and Travelism: Letters from Leaders*, edited by G. Lipman, T. Delacy, S. Vorster, R. Hawkins and M. Jiang, 193–197. Oxford: Goodfellow Publishers, 2012.

Death, C. "The Green Economy in South Africa: Global Discourses and Local Politics." *Politikon* 41, no. 1 (2014): 1–22.

Dodds, F. "Traveling to Tomorrow – Stakeholders on the Same Planet." In *Green Growth and Travelism: Letters from Leaders*, edited by G. Lipman, T. Delacy, S. Vorster, R. Hawkins and M. Jiang, 284–289. Oxford: Goodfellow Publishers, 2012.

Dryzek, J. S. *The Politics of the Earth: Environmental Discourses*. 3rd ed. Oxford: Oxford University Press, 2013.

Dubois, G., P. Peeters, J. P. Ceron, and S. Gössling. "The Future Tourism Mobility of the World Population: Emission Growth versus Climate Policy." *Transportation Research Part A: Policy and Practice* 45, no. 10 (2011): 1031–1042.

Duffy, R. *Nature Crime: How we're getting Conservation Wrong*. New Haven, CT: Yale University Press, 2010.

Ehresman, T. G., and C. Okereke. "Environmental Justice and Conceptions of the Green Economy." *International Environmental Agreements* 15, no. 1 (2015): 13–27.

Enders, T. "One Born Every Minute." In *Green Growth and Travelism: Letters from Leaders*, edited by G. Lipman, T. Delacy, S. Vorster, R. Hawkins and M. Jiang, 27–32. Oxford: Goodfellow Publishers, 2012.

European Commission. "Communication from the Commission to the European Parliament, the Council, the European Economic and Social Committee and the Committee of the Regions: a roadmap for moving to a competitive low carbon economy in 2050." Brussels, 2011. http://eur-lex.europa.eu/legal-content/EN/TXT/PDF/?uri=CELEX:52011DC0112&from=EN.

Francis, P. R. "Greening the Tourism Sector: Building the Competitiveness of Developing Countries." In *Green Growth and Travelism: Letters from Leaders*, edited by G. Lipman, T. Delacy, S. Vorster, R. Hawkins and M. Jiang, 33–39. Oxford: Goodfellow Publishers, 2012.

Gittens, A. "Aviation and Green Growth – An Airport Perspective." In *Green Growth and Travelism: Letters from Leaders*, edited by G. Lipman, T. Delacy, S. Vorster, R. Hawkins and M. Jiang, 43–51. Oxford: Goodfellow Publishers, 2012.

Gössling, S., C. M. Hall, and D. Scott. "The Challenges of Tourism as a Development Strategy in an Era of Global Climate Change." In *Rethinking Development in a Carbon-constrained World*, edited by Eija Palosuo, 100–119. Helsinki: Ministry of Foreign Affairs, 2009.

Gössling, S., C. M. Hall, P. Peeters, and D. Scott. "The Future of Tourism: Can Tourism Growth and Climate Policy be Reconciled? A Mitigation Perspective." *Tourism Recreation Research* 35, no. 2 (2010): 119–130.

Gössling, S., and P. Peeters. "'It Does Not Harm the Environment!' An Analysis of Industry Discourses on Tourism, Air Travel and the Environment." *Journal of Sustainable Tourism* 15, no. 4 (2007): 402–417.

Gramsci, A. *Selections from the Prison Ntebooks*. New York: International Publishers, 1971.

Guevara, G. "Tourism, Green Growth and Sustainable Development." In *Green Growth and Travelism: Letters from Leaders*, edited by G. Lipman, T. Delacy, S. Vorster, R. Hawkins and M. Jiang, 56–59. Oxford: Goodfellow Publishers, 2012.

Hall, C. M. "Degrowing Tourism: Décroissance, Sustainable Consumption and Steady-state Tourism". Anatolia: An International Journal of." *Tourism and Hospitality Research* 20, no. 1 (2009): 46–61.

Hall, C. M. "Framing Behavioural Approaches to Understanding and Governing Sustainable Tourism Consumption: Beyond Neoliberalism, 'Nudging' and 'Green Growth'?" *Journal of Sustainable Tourism* 21, no. 7 (2013): 1091–1109.

Hall, C. M., D. Scott, and S. Gössling. "The Primacy of Climate Change for Sustainable International Tourism." *Sustainable Development* 21, no. 2 (2013): 112–121.

Harcourt, W. "Editorial: The Times they are A-changin'." *Development* 55, no. 1 (2012): 1–4.

Hogan, J. "Green Growth and Travelism." In *Green Growth and Travelism: Letters from Leaders*, edited by G. Lipman, T. Delacy, S. Vorster, R. Hawkins and M. Jiang, 70–74. Oxford: Goodfellow Publishers, 2012.

IPCC. *IPCC Fourth Assessment Report: Climate Change 2007, Working Group III Report 'Mitigation of Climate Change'*. 2007. http://www.ipcc.ch/pdf/assessment-report/ar4/wg3/ar4_wg3_full_report.pdf.

Jessop, B. "Economic and Ecological Crises: Green New Deals and No-growth Economies." *Development* 55, no. 1 (2012): 17–24.

Kallis, G. "In Defence of Degrowth." *Ecological Economics* 70, no. 5 (2011): 873–880.

Kosoy, N., P. G. Brown, K. Bosselmann, A. Duraiappah, B. Mackey, J. Martinez-Alier, D. Rogers, and R. Thomson. "Pillars for a Flourishing Earth: Planetary Boundaries, Economic Growth Delusion and Green Economy." *Current Opinion in Environmental Sustainability* 4, no. 1 (2012): 74–79.

Lee, D. S., D. W. Fahey, P. M. Forster, P. J. Newton, R. C. N. Wit, L. I. Lim, B. Owen, and R. Sausen. "Aviation and Global Climate Change in the 21st Century." *Atmospheric Environment* 43, no. 22 (2009): 3520–3537.

Lee, J. J. "Can we Accelerate the Improvement of Energy Efficiency in Aircraft Systems?" *Energy Conservation and Management* 51, no. 1 (2010): 189–196.

Lefebvre, M. "The Green Economy – Travel and Tourism and the Cruise Sector." In *Green Growth and Travelism: Letters from Leaders*, edited by G. Lipman, T. Delacy, S. Vorster, R. Hawkins and M. Jiang, 94–99. Oxford: Goodfellow Publishers, 2012.

Lipman, G., T. DeLacy, S. Vorster, R. Hawkins, and M. Jiang. "Introduction and Overview." In *Green Growth and Travelism: Letters from Leaders*, edited by G. Lipman, T. Delacy, S. Vorster, R. Hawkins and M. Jiang, 1–7. Oxford: Goodfellow Publishers, 2012.

Lipman, G., T. DeLacy, S. Vorster, R. Hawkins, and M. Jiang. "Towards Transformation." In *Green Growth and Travelism: Letters from Leaders*, edited by G. Lipman, T. Delacy, S. Vorster, R. Hawkins and M. Jiang, 8–21. Oxford: Goodfellow Publishers, 2012.

Lipman, G., T. DeLacy, S. Vorster, R. Hawkins, and M. Jiang, eds. *Green Growth and Travelism*. Oxford: Goodfellow Publishers.

Lyle, C. "Aviation's Role in Green Growth for Developing Countries." In *Green Growth and Travelism: Letters from Leaders*, edited by G. Lipman, T. Delacy, S. Vorster, R. Hawkins and M. Jiang, 100–105. Oxford: Goodfellow Publishers, 2012.

Meadows, D. H., and D. L. J. Meadows. *The Limits to Growth*. New York, NY: New American Library, 1972.

Newsome, D., S. A. Moore, and R. K. Dowling. *Natural Area Tourism: Ecology, Impacts and Management*. Clevedon: Channel View Publications, 2002.

Oppermann, M. "Tourism Space in Developing Countries." *Annals of Tourism Research* 20, no. 3 (1993): 535–556.

Panitchpakdi, S. "Sustainable Tourism Development – A Public-Private Partnership." In *Green Growth and Travelism: Letters from Leaders*, edited by G. Lipman, T. Delacy, S. Vorster, R. Hawkins and M. Jiang, 112–117. Oxford: Goodfellow Publishers, 2012.

Pearce, D. W., A. Markandya, and E. Barbier. *Blueprint for a Green Economy*. London: Earthscan, 1989.

Peeters, P. M., and E. Eijgelaar. "Tourism's Climate Mitigation Dilemma: Flying between Rich and Poor Countries." *Tourism Management* 40 (2014): 15–26.

Peeters, P., and G. Dubois. "Tourism Travel under Climate Change Mitigation Constraints." *Journal of Transport Geography* 18, no. 3 (2010): 447–457.

Rifai, T. "One Billion Tourists…One Billion Opportunities for more Sustainable Tourism." In *Green Growth and Travelism: Letters from Leaders*, edited by G. Lipman, T. Delacy, S. Vorster, R. Hawkins and M. Jiang, 128–133. Oxford: Goodfellow Publishers, 2012.

Saarinen, J., C. Rogerson, and H. Manwa. "Tourism and Millennium Development Goals: Tourism for Global Development?" *Current Issues in Tourism* 14, no. 3 (2011): 201–203.

Sandbrook, C. G. "Putting Leakage in its Place: The Significance of Retained Tourism Revenue in the Local Context in Rural Uganda." *Journal of International Development* 22, no. 1 (2010): 124–136.

Scheyvens, R. "Ecotourism and the Empowerment of Local Communities." *Tourism Management* 20, no. 2 (1999): 245–249.

Schilcher, D. "Growth versus Equity: The Continuum of Pro-poor Tourism and Neoliberal Governance." *Current Issues in Tourism* 10, nos. 2–3 (2007): 166–193.

Scott, D., P. Peeters, and S. Gössling. "Can Tourism deliver its 'Aspirational' Greenhouse Gas Emission Reduction Targets?" *Journal of Sustainable Tourism* 18, no. 3 (2010): 393–408.

Scowsill, D. P. "Travel and Tourism leading the Way towards Green Growth." In *Green Growth and Travelism: Letters from Leaders*, edited by G. Lipman, T. Delacy, S. Vorster, R. Hawkins and M. Jiang, 152–155. Oxford: Goodfellow Publishers, 2012.

Sharpley, R. *Tourism Development and the Environment: Beyond Sustainability?*. London: Earthscan, 2009.

Tienhaara, K. "Varieties of Green Capitalism: Economy and Environment in the Wake of the Global Financial Crisis." *Environmental Politics* 23, no. 2 (2014): 187–204.

Timilsina, G. R., and A. Shrestha. "How much Hope should we have for Biofuels?" *Energy* 36, no. 4 (2011): 2055–2069.

Tyler, T. "A Green Economy: The Contribution of Travel and Tourism – The Aviation Viewpoint." In *Green Growth and Travelism: Letters from Leaders*, edited by G. Lipman, T. Delacy, S. Vorster, R. Hawkins and M. Jiang, 179–185. Oxford: Goodfellow Publishers, 2012.

UNCTAD. *Sustainable Tourism: Contribution to Economic Growth and Sustainable Development*. 2013. www.unctad.org/meetings/en/SessionalDocuments/ciem5d2_en.pdf.

UNEP. *Green Economy: Developing Countries Success Stories*. 2010. www.unep.org/pdf/greeneconomy_successstories.pdf.

UNEP. *Towards a Green Economy: Pathways to Sustainable Development and Poverty Eradication*. 2011. www.unep.org/greeneconomy.

UNWTO. *From Davos to Copenhagen and Beyond: Advancing Tourism's Response to Climate Change – UNWTO Background Paper*. 2009. http://sdt.unwto.org/sites/all/files/pdf/537_from_davos_to_copenhagen_and_beyond_unwto_paper_electronic-version_lr.pdf.

UNWTO. *UNWTO Tourism Highlights*. 2014. http://dtxtq4w60xqpw.cloudfront.net/sites/all/files/pdf/unwto_highlights14_en.pdf.

UNWTO, European Travel Commission and VisitSweden. *Travel and Tourism in the Green Economy: Symposium Conclusions*. 2009. http://sdt.unwto.org/event/symposium-tourism-travel-green-economy.

UNWTO, UNEP, and WMO. *Climate Change and Tourism: Responding to Global Challenges*. 2008. http://sdt.unwto.org/sites/all/files/docpdf/climate2008.pdf.

van Schalkwyk, M. "Breaking out of the Silos." In *Green Growth and Travelism: Letters from Leaders*, edited by G. Lipman, T. Delacy, S. Vorster, R. Hawkins and M. Jiang, 186–192. Oxford: Goodfellow Publishers, 2012.

Wanner, T. "The New 'Passive Revolution' of the Green Economy and Growth Discourse: Maintaining the 'Sustainable Development' of Neoliberal Capitalism." *New Political Economy* 20, no. 1 (2015): 21–41.

Wapner, P. "Sacrifice." In *Critical Environmental Politics*, edited by C. Death, 208–217. Abingdon: Routledge, 2014.

Suspended redistribution: 'green economy' and water inequality in the Waterberg, South Africa

Michela Marcatelli

International Institute of Social Studies, Erasmus University Rotterdam, The Hague, The Netherlands

In this article I show how ideas and practices of 'green economy' can reproduce and even naturalise inequality in water access for local users. Evidence to support my argument is drawn from the Waterberg region in the Limpopo Province of South Africa. Following the demise of apartheid and the appeal of the green economy, the Waterberg has been 'reinvented' as a wildlife destination. Whereas game farms enjoy secure water supply, the rural poor relocated to the small town of Vaalwater suffer severe water shortages. The article questions the mainstream view according to which game farms have no relationship to the water problems in town. Rather, I suggest that by conceiving and managing water as a private commodity deriving from land ownership and largely unregulated by the state, green economy initiatives contribute both materially and discursively to hampering more equality in water redistribution.

Introduction

The Waterberg Biosphere Reserve is a magical part of South Africa which is easily accessible from Africa's industrial powerhouse, Gauteng. It is very old, and yet a very new place too. With its unique history of sparse human settlement, it has been perfectly placed to reinvent itself, following the dawn of democracy in South Africa, as a stunningly beautiful and highly significant conservation area.[1]

These few opening lines of an attractive brochure aimed at guiding tourists through the meanders of the Waterberg plateau dirt roads perfectly sketches the contours of the myth on which the production of this area as a wildlife destination rests. Although it provides traces of human presence from thousands of years ago, it was only in the second half of the 19th century that a handful of white settlers occupied this part of Northern Transvaal (today Limpopo) permanently. According to the myth, this 'unique' population dynamic – partly explained by the difficult environmental conditions of the plateau, bounded by the Waterberg escarpment – contributed to the preservation of the place as an 'unspoilt wilderness' and adequately demonstrates why it was 'perfectly placed' to respond to the appeal of the 'green economy' mostly via private conservation activities, such as game farms and nature reserves.[2] The advertising board at the entrance of the local Spar invites visitors to 'Relax. You're in The Bush.' Surely, this is what Pretoria and Johannesburg residents, who stop in the small town of Vaalwater after a three-hour drive to buy the last supplies for a weekend getaway – but also international tourists who finally stretch their legs after having been picked up at OR Tambo International Airport – often do. In order to reach their destinations (be it a luxurious lodge or a self-catering cottage), they do not need to drive around the dusty and crowded streets of the local township of Leseding. By staring at the vast stretches of hilly bushveld landscape – always fenced – outside their car windows and making a couple of game sightings, their dream of being at one with nature is enacted.

Since the very purpose of this brochure and other popular publications, which have recently appeared on the Waterberg,[3] is to forge and convey a simple story to tourists, they tend to overlook possible elements of tension, such as the fact that most of the game that can be admired in the region today was actually reintroduced over the past 30 years. Above all, the myth about the Waterberg being essentially a conservation area tends to disembed it from the agrarian political economy of the place as shaped by colonialism and apartheid. By putting conservation back into its agrarian context, we can see how it influences the redistribution of natural resources and determines who is winning and who is losing from these green development processes.

In this article I focus on water to show how specific ideas and practices of green economy can reproduce and even naturalise inequality in access for local users. Whereas on private nature reserves and game farms water *must* be abundant in order to guarantee that the demands of landowners, tourists and wildlife are satisfied, water provision in Vaalwater *can* be interrupted to the point where the minimum standards for basic water provision are not met. This situation is especially affecting the black and poor population of Leseding, where people are progressively relocating from private farms and the rural villages in the former Bantustan of Lebowa.

The article questions the mainstream view according to which game farms are saving water and do not have a relationship with the water problems experienced in town. Rather, I suggest that they do consume water and contribute both materially and discursively to hamper more equality in water redistribution. More specifically I identify three ways in which private nature conservation is actively reproducing water inequality in the Waterberg: by conceiving and managing water as a commodity deriving from land ownership: by being excepted from state regulation; and by excluding local black people (except as labour) from its project of social production of space.

The quotation opening this section argues for a compelling linkage between democracy, transformation ('reinvention') and nature conservation. From a water political economy perspective, however, the transformation taking place in the Waterberg appears to be fundamentally conservative, insomuch as local power relations remain unchallenged and private control over natural resources (especially land and water) is tightened. It is here that I see the main reason why water redistribution is currently failing in the plateau, as opposed to the lack of municipal capacity argument dominating the country's public discourse both at the national and local level.

The data presented in the article were collected during a one-year period of fieldwork in the Waterberg plateau, carried out between August 2013 and August 2014. I engaged with Vaalwater and Leseding residents and with owners and managers of private nature reserves and game farms through a combination of social research methods.

The rest of the article is structured in three major sections. First, I situate my contribution within the debates on the changing agrarian political economy of South Africa and the social consequences of private conservation initiatives. Second, I briefly sketch the history of the production of the Waterberg as a con-servation site and point out some recent trends in this on-going process. Finally, I analyse water inequality in the Waterberg by considering the specific ways in which game farms and private nature reserves reproduce and naturalise it.

Situating water within agrarian political economy debates

The Waterberg falls within the borders of former *white* rural South Africa, where the land-water nexus is paramount. Being dependent upon access to land, the distribution of water resources is similarly deeply unequal and skewed along racial lines.[4] For this reason it is crucial to situate our discussion on water access within a broader debate on the agrarian political economy of the place, revolving around issues of land, labour and livelihoods. In this way I also intend to problematise a tendency in South African public discourse towards establish-ing dualisms – in this context, farm vs town(ship) – by unravelling the historical and present-day relationships between people and spaces.[5]

One of the most important lines of enquiry to recently emerge in the field of critical agrarian studies refers to the new phenomenon of land and resource accumulation known as 'grabbing'.[6] In the course of the years authors have moved beyond a focus on agriculture alone as the main driver of land grabs to conceptualise the notions of 'green grabbing' and 'water grabbing'.[7] Green grab-bing has been defined as a dynamic of accumulation (by powerful actors) and dispossession (of poor and marginalised communities) for declared environmen-tal purposes. What is qualitatively new about this process is that it takes place in a context where environmental concerns have become mainstream – the very notion of green economy being a case in point – and nature is commodified to provide new avenues both for capital accumulation *and* to 'repair' environmen-tal loss (in line with the notion of neoliberal conservation).[8] Water grabbing, on the other hand, has been identified as an issue apart, to emphasise the fact that, without secure access to water, agricultural land has no value but also to point out that water itself can be the object of grabbing (especially in relation to hydropower and mining projects).[9] Water is fluid in time and space, however, and this makes it more difficult to reallocate control over it as well as to evalu-ate the social consequences of the grabbing.

The scholarship on land and resource grabbing cautions us that, although these are to be interpreted as global phenomena related to the contemporary phase of neoliberal capitalism, local contexts always matter in that it is they that will ultimately shape the specific forms in which the (re)appropriation takes place. Moreover, to fully grasp the sense of injustice that is conveyed by the expression 'grabbing', it is important to keep in mind the particular histories of dispossession that characterise a specific place. For this, we need to turn our attention to the agrarian question in post-apartheid South Africa.

Hitherto South African land reform has had meagre and even controversial results, with the deadline for redistributing 30% of the land postponed to 2025, thousands of restitution claims yet to be settled or recently reopened and almost 1.7 million people evicted from commercial farms between 1984 and 2004.[10] Yet it is so full of symbolic meanings (such as restorative justice and national identity) that it cannot simply be put aside.[11] It is not all about land, however. The contemporary agrarian question has been rephrased to ask whether and how a comprehensive agrarian reform can contribute to solving the extreme poverty (both in rural areas and in slum settlements around cities) and growing inequality characterising the country.[12]

The conversion of commercial farms from traditional activities like crop and livestock production to wildlife production (eco-tourism, hunting, venison production, game breeding and trading) represents an important land-use change in the agrarian landscape that has prompted the emergence of a critical scholarship in recent years.[13] Some authors do not hesitate to interpret farm conversion as a local manifestation of green grabbing on the basis of the transformation of wildlife into a commodity, whose value has been escalating, and of the fact that only the wealthy can afford to buy game and the large tracts of wilderness it needs.[14] This new practice of land enclosure does indeed seem at odds with the purpose of justice embedded in the land reform and research has shown how nature conservation can actually work as a strong moral justification to keep both the government and claimants at bay, at the same time helping white farmers negotiate their new role in democratic South Africa.[15] Other authors, however, have nuanced the discursive and material contours of the dispossession suffered by the people who live and work the land without owning it (ie farm workers and dwellers) following the conversion to game farming.[16] Their particular histories of past displacement and mobility seem in fact to account for whether the conversion is perceived as yet another round of exclusion or as a decisive rupture.

The lives and experiences of the rural 'working poor' thus come to the fore when we put our analysis of the green economy into agrarian contexts.[17] Existing research on private game farming has indeed focused on the social consequences that conversion entails for these subjects, especially their tenure and labour relations. Scholars have found that, contrary to mainstream views about the positive contribution of private conservation to poverty alleviation,[18] game farms offer fewer job opportunities than traditional ones, the positions offered are usually low-skilled, and salaries are aligned to those employed in agriculture, which are the lowest.[19] Moreover, the presence of fences and dangerous game affects people's mobility and their ability to keep livestock and access grazing. The loss of jobs coupled with increasingly difficult living conditions has caused many workers and dwellers to leave the farm voluntarily or forcibly, thus fulfilling the idea that wild nature must be emptied of human presence (or at least of some humans).[20] Displaced workers make for a new group of 'surplus people' with no other choice than relocating to informal settlements or rural towns.[21] It is at this juncture that a water perspective can provide new insight into the debates discussed above. Before addressing water issues in more depth, however, the next section provides an overview of the process of conversion that has been taking place in the Waterberg.

Producing nature in the Waterberg: from traditional to game farming

Similarly to the rest of South Africa the first private nature reserves were proclaimed by Waterberg landowners in the 1960s.[22] However, conservation activities became common only in the 1980s, following the initiative of some wealthy white businessmen and farmers, usually self-proclaimed 'conservationists' who intended to bring the place back to its 'original wilderness'. As one of the landowners that I interviewed put it, she started the reserve to rescue a land 'destroyed' by overgrazing and to bring the bush back to its 'natural' status.[23] At that time the landscape consisted mainly of livestock and crop private farmland. Cattle farming was the most widespread land use because of the dry climate and a soil poor in nutrients, alternating rocks and sand. Yet, where irrigation was possible, that is, along the Mokolo River, sandy soils favoured the development of tobacco farming, making Vaalwater one of the major tobacco growing areas of the country to date.

Within a context of deregulation of agriculture – that is, the removal of marketing boards and other state subsidies, which started in the mid-1970s and accelerated after 1994[24] – and following the institution of private ownership of game (via the Game Theft Act, 1991), the conversion of traditional farmland into private nature reserves and game farms started to make economic sense.[25] Only a few landowners were able to fund the conversion, however. The development of Welgevonden Game Reserve, for instance, albeit initiated in 1987 by the farm owner, Pienkes Du Plessis, was soon taken over by Rand Merchant Bank. Most of the time it was wealthy individuals from other parts of the country, or even from abroad, who bought the land from local farmers in financial difficulties and then invested their own capital to incorporate more land from adjoining farms, bring down cattle fences and other farming infrastructure, introduce game – long disappeared thanks to hunting for trade in the Transvaal of the 19th century and agriculture thereafter – and to build suitable fences.

Although some of the new owners employed their game-stocked properties as family hunting farms, a preservationist approach seemed to build momentum in the 1980s. This was largely promoted by the figure of Clive Walker, game ranger, artist and founder of the Endangered Wildlife Foundation, who moved to the plateau around that time. In 1981 Walker found in the businessman Dale Parker an investor for the purchase of a farm, which was later developed into the 36,000 ha private nature reserve Lapalala Wilderness. Similarly Walker was able to reach other 'like-minded' people with the means to buy land adjoining Lapalala. Then, in 1990, he prompted the foundation of the Waterberg Nature Conservancy, whose first members were Lapalala and the two neighbouring game farms Kwalata Wilderness and Touchstone Game Ranch. The original scope of the Conservancy was to take all the fences down and transform the Waterberg into an extensive wilderness, with no human use allowed. This triggered opposition from the farming community and the establishment of a frontline between farmers (mostly Afrikaners) and conservationists (mostly English-speaking) which, to a large extent, continues to date. Conservationists won an important battle at the time, as Lapalala disputed and eventually halted the construction of a government dam on its land – a so-called 'election dam', meant to secure farmers' votes – to protect the 'pristine' Lephalala River system.[26]

Notwithstanding the Conservancy's original purposes, more fences have actually gone up in the course of the years and conservation in the plateau is now largely managed according to commercial principles. Under the ownership of Duncan Parker (son of Dale), even Lapalala has started a partnership with businessmen Gianni Ravazzotti and Peter Anderson; the three of them have developed a 'bold' plan to assure more funding for the reserve activities.[27] The plan includes the following: expanding the already existing special species breeding project; enhancing tourism; and offering to individuals and companies the opportunity to invest in the reserve and become 'custodians' of the land.

Local authorities are supportive of this shift in the modes of production from traditional agriculture to green activities and intend to sustain it by developing a 'Waterberg brand' that would make the place distinctive and competitive on the global tourism market.[28] The proclamation of the Waterberg Biosphere Reserve by UNESCO in 2001 also contributed to increasing the international visibility of the Waterberg, although some locals remain sceptical of the work done on the ground by the NGO administering it.[29]

To protect nature while at the same time making a profit out of it has meant focusing on those conservation activities that can guarantee the highest returns to landowners. In a relatively small area such as the plateau, with a billboard advertising a game lodge at every turn of the (dirt) road, this implies catering for the needs of upper market eco-tourists and overseas hunters, who are willing to spend up to around ZAR5000 per day – or even ZAR10,000 in the most exclusive of lodges.[30]

Besides eco-tourism and trophy hunting, a sub-sector, which has gained prominence in the past 10 years, is that of game breeding. Snijders has documented the 'escalation of commodity value' of wildlife, whose turnover increased from ZAR9 million in 1991 to ZAR303 million in 2010.[31] More recently the Deputy President of South Africa, Cyril Ramaphosa, who owns a game farm in the district, hit the headlines for bidding ZAR19.5 million for a buffalo cow.[32] Although some insiders do not hesitate to qualify it as a market bubble, ready to burst at any time, investors keep trading in live game (especially rare and exotic species), attracted by a return of 300% or even 400%.[33] Small ranchers, who start afresh in the wildlife industry, keep the demand for common game high (sometimes supplied by livestock farmers diversifying their activities), whereas big players provide the individual or corporate capital necessary to specialise in genetics and to produce game of higher value. Apart from meat production, which is partly for export, the end-uses of game are mainly local: wildlife is sold to other farms for trophy and recreational hunting, safaris and further breeding. On the Waterberg plateau the two reserves which have made breeding their core business and are now in a position to organise their own auctions are Keta Private Game Reserve and Shambala Game Reserve, both owned by white South African millionaires.[34] Keta alone made a profit of

ZAR26 million at its last auction in May 2014.[35] Not only have game auctions surpassed cattle ones in frequency and sale volume, but they have also become an important social event for Waterberg landowners. One can immediately recognise an auction day by the unusually high number of cars – plus a few helicopters – parked on the side of the tar road. In principle, all local families can afford the opportunity to admire fancy animals, such as black impalas and golden gnus – kept in pens though, not roaming 'freely' in the bush; nevertheless, apart from black workers, white and khaki dominate the landscape.

Another trend, which has (re)emerged in the past 10 years, is the increase in the number of reserves and farms for the private use of landowners, sometimes in the form of wildlife estates. Here, the commodification of wilderness occurs through the valuation of converted land, which creates new investment opportunities for the well-to-do. There may be situations where a landowner rents out a small cottage to weekend tourists or starts a breeding project, but only to make an extra income, since he or she does not need to make a living out of the land. In the Waterberg, those who own a game farm or a portion of a wildlife estate for the purpose of spending weekends, enjoying an early retirement or even starting a family in close contact with 'nature' constitute a diversified group.[36] Generally speaking, however, they tend to be white and use English as a medium of communication. The closeness (in terms of South African distances) of the plateau to Johannesburg international airport has turned the place into a haven for foreigners eager to buy their own 'piece of paradise' and prompted by a favourable exchange rate. The majority of foreign landowners come from Europe, but some travel from as far as the USA to spend a week or two every year in their bush home. Besides practical considerations, such as its being close to Gauteng, malaria-free and much cheaper than the Cape, what makes the plateau attractive to potential overseas buyers is that it matches quite well their ideas of wild Africa as an 'empty' land. This is not to say that there are no people living in the area, of course, but as long as white people stay on their secluded farms and black people are gathered in one *location* – instead of being scattered all over the place – the illusion is preserved. Moreover, this contributes to the perception of a safe countryside compared with other parts of South Africa.[37]

A useful indicator of the upsurge in the demand for game farms as private residential land is the composition of the Waterberg Nature Conservancy membership. The number of Conservancy members who own land for private use only doubled between 2002 and 2010, amounting to 16, or 40% of the total.[38] This change is reflected in the Conservancy activities – often described by non-members as an 'exclusive, wealthy, English-speaking club'[39] – now revolving around a general meeting held once every two months where, in addition to housekeeping matters, a guest speaker gives a talk on something broadly related to environmental conservation.

Eventually the uplift of the local community (read blacks) has made it on to the agenda of the Conservancy and of other green economy businesses in the Waterberg. Besides the usual rhetoric of helping the poor by creating new jobs – particularly questionable in the case of game breeding and residential developments – this has implied the establishment of a few charities.[40] It is not to deny the good done by such initiatives to note that, by giving back just a little to the community, without putting its 'poverty and disease' in relation to the history and political economy of the place,[41] the unequal distribution of resources in the area never comes into question and ends up being reinforced.[42] Access to water offers a clear example of this.

Following water through farms and town

During the colonial and apartheid times the Waterberg's landscape was actively produced to accommodate the needs of a small class of white commercial farmers. Its water resources (both surface and underground) were therefore appropriated by landowners, who used them according to the following hierarchy: their produce, their own reproduction, and that of their workers. The small town of Vaalwater was founded at the beginning of the 20th century as a service point for the local farming community. Following the 1931 Transvaal Townships and Town-planning Ordinance, based on the principle of racial segregation, urban planning provided for water being supplied from the nearby Mokolo River, as long as this would not undermine the irrigation rights of riparian farms.[43]

Plans for the establishment of a township for 'non-Europeans' were discussed in 1948, but nothing was implemented; since Vaalwater was located in the midst of European farmers, the landscape was to be cleared of black presence too. In 1965 the town was declared a 'whites-only' area on the basis of the Group Areas Act, 1950 and black residents were displaced to the Bantustan of Lebowa.[44] It was only in 1996 that the newly elected African National Congress (ANC) government authorised the foundation of the township of Leseding, by providing the first RDP (state-subsidised) houses. Some blacks relocated there from white farms because they were tired of living on someone else's land, prone to their *baas*'s (master) abuses, whereas for those living in the rural villages of Lebowa the township offered more opportunities in terms of jobs and services (such as schools, shops and the clinic). However, many did not choose to live in Leseding, they simply did not have any alternative. Feeling threatened by the perspective of an imminent land reform, some white farmers loaded their trucks with their workers' families and moved them to the township. Other black families living on white land were evicted when the elderly – the only ones who were still working the land – could not work anymore or when the property was sold and converted into a game farm.

As those Leseding residents who were children in the mid-1990s started to have their own families, and foreign migrants (from Zimbabwe and Mozambique), who became instrumental in the economy of the region as a pool of cheap labour on crop and cattle farms, started to arrive, the demand for water services in town increasingly exceeded the supply, to the point that today residents suffer from severe water shortages.[45] Table 1 provides a description of water access in the Waterberg, by considering how many hours per day water is actually running through taps and how many litres of water per day a person is able to consume for domestic uses. The data show how water access is highly unequal and ultimately dependent upon settlement patterns, namely whether one resides in the former white suburbs, in the township, or on a private game farm.

Two issues in relation to the data need to be clarified. First, the striking difference in water access between suburbs and township (both served by Modimolle Local Municipality (MLM)[46]) is explained by better water infrastructure in the suburbs coupled with many of their residents having the means to drill

Table 1. Access to water in the Waterberg.

Typology of settlement	Water availability (hrs/d)	Water consumption (l/c/d)
Vaalwater suburbs	9	195
Leseding Ext #1	2	26
Leseding Ext #2	2	46
Leseding Ext #3	3	32
Leseding Ext #4	4	26
Leseding Ext #5	4	27
Leseding Ext #6	3	22
Game farms/private nature reserves	24	606

Source: data from structured and semi-structured interviews with 90 residents.

a borehole in their yard or buy a 15,000-litre water tank. Second, the even more striking difference between town and game farms is explained in part by the fact that trying to focus only on the domestic uses of water on farms, for the purpose of comparing them with those in town, turns out to be an almost impossible task, because water consumption on the former is rarely metered and the same source is normally used for several activities at the same time. Moreover, water infrastructure serving the needs of owners, managers and tourists can be very different from that supplying farm workers.

If coping with water shortages in town may ultimately be seen as a matter of individual capacity of storing water for personal consumption, then in Leseding residents can rarely afford the luxury of a water tank and have to rely on more mundane containers, hence the far inferior amounts of water they are able to consume. The dependence on containers is already evident in Extension 1 and 2, where RDP houses are provided with in-house and yard taps, respectively and residents have to be ready to collect water whenever it comes, but it becomes all the more urgent in the other four extensions of the township, which are provided only with communal taps. Here, access to water actually means waking up at 4 am or 5 am to start queuing at the tap, hoping that water will last until your turn comes.

The mainstream narrative among those who do not live in town and do have secure water access maintains that such problems are the direct result of the municipality's lack of capacity coupled with continuous influxes of illegals 'who were never meant to be there'.[47] However, my argument is that a major redistribution of people – mainly black farm workers becoming superfluous to the economic needs of white farming *and* inimical to the whites' politics of place based on nature conservation – has not been followed by an adequate redistribution of resources that would satisfy basic human needs, such as safe and continuous water access. The sense of injustice conveyed by this – and seen in the light of the history of dispossession of the place – is clearly not shared by everyone. In the imagination of white landowners the fences demarcating their private property act as a border separating the world of the farm from that of town. Anything happening outside a private fence automatically becomes the municipality's responsibility. No relationships are drawn between worlds so far apart, except an emotional response (translating into charity initiatives) to the stark contrast between abundance and deprivation characterising the place. Instead, I contend that such relationships between people and space exist and it is important to trace them, hence to follow water through farms and town.

From the perspective of the local municipality, water supply in Vaalwater cannot be increased for two main reasons. First, water sources are to be found on private land and landowners – especially irrigation farmers – fight hard to keep control over their 'existing lawful uses'.[48] For instance, although there are landowners whose water allocation exceeds their current needs, and who are keen on making a profit by renting their boreholes to the municipality, they may face opposition from other farmers, who think their own activities will be put at risk.[49] Second, a large part of the water resources of the plateau, collected in the Mokolo Dam (about 60 km to the north of Vaalwater), have been earmarked for other national 'strategic' uses. In fact, this dam is intended to supply water to the new Eskom Medupi Power Station, a massive coal plant near the town of Lephalale, which is expected to solve the country's energy crisis.[50]

Up to now private nature conservation has tended to remain in the background of the South African water debate. I think it is time to bring this important land-use change and its relationship to water resource to the fore. I will start by looking at three specific ways in which game farms actively reproduce water inequality in the Waterberg both materially and discursively.

First, in the National Water Act, 1998 (NWA) – the cornerstone of the post-apartheid water legislative framework – water is declared 'a natural resource that belongs to all people' under public trusteeship and the notion of water *rights* is replaced with that of water *uses*, which need to be regulated by the state via registration and licensing.[51] However, game farmers – similarly to traditional farmers – do not conceive water as a public or common good, which can be redistributed to accommodate the needs of all, but rather as a private one that they rightfully appropriated when purchasing the land. The fact that they do not receive a service from the municipality, but have to provide water for themselves, by pumping it out of a river or a borehole, reinforces their perception of private ownership over this natural resource. Since water is 'theirs', game farmers also feel that they can legitimately do anything they want with it. For instance, they can prevent any extraction from 88 km of river shoreline for conservation purposes, such as in the case of Lapalala Wilderness. Most of the time, however, they do extract water and, especially when offering eco-tourism services, this goes well beyond the satisfaction of basic needs. A 'wild' experience in the bush in fact seems inconceivable without running taps, toilets connected to a sewage system (mostly a septic tank) and amenities such as swimming pools, Jacuzzis and private dams.

Second, given the Department of Water and Sanitation's (DWS) narrow focus on irrigation,[52] private conservation activities have not been targeted for the purpose of regulating water uses and redressing past inequalities in water allocation. As a result, game farms and private nature reserves know very little about Water Use Licensing, Registration and Revenue Collection (WARMS) and usually do not register their water use, apply for a licence or pay water fees. Two narratives are employed by game farmers to justify their exemption.

On the one side, game farmers claim to use little water – that is, in comparison to irrigation farmers. However, most of them do not monitor their water consumption and therefore cannot support their claims with actual figures. They do not see the point of installing a water meter, since they have water in abundance and do not need to pay for what they consume. In my research I tried together with game farm owners and managers to calculate – usually on the basis of educated guesses – what their average daily water consumption was. The extreme results are quite interesting: a small (3000 ha) farm with three permanent residents (two owners and one staff member) and the capacity to accommodate up to 12 guests (generally over weekends) would consume around 1000 l/d, whereas a big (34,000 ha) reserve with 350 permanent residents on average (guests and staff) would consume around 275,000 l/d.[53] A number of factors contribute to explain this difference, namely: farm size; the number of permanent residents; the game species present (and whether the farmer waters them during the dry winter months); and the types of activity conducted (whether the farmer grows lucerne to feed the game, irrigates lawns, offers tourists horse riding safaris or even the possibility of playing golf). The point here is that water does turn out to be a strategic resource for game farms, too. Indeed, checking the availability of water sources before the purchase of a property is as important as checking for possible land claims. Nonetheless, there is a serious lack of data about the quantities of water actually consumed. Furthermore, in the absence of a clear definition of 'small volumes' – the threshold for registration and licensing, according to DWS – the initiative to approach the Department is left to the discretion of individual landowners, who usually do not want any government interference in their activities for fear of losing what they perceive as 'their' water and ending up paying more taxes.

On the other side, game farmers maintain that – again, since they do not irrigate – they employ water only for *domestic* purposes and therefore fall within the category of Schedule 1 water uses, that is, permissible uses according to the NWA. Nevertheless, the Act reads 'A person may, subject to this Act take water for reasonable domestic use in that person's household, directly from any water resource to which that person has lawful access',[54] whereas on game farms water often becomes an essential component of a commercial service (think of eco-tourism or game trading) and therefore in need of authorisation.

Finally, not only does the conversion to game farming *de facto* reproduce a system whereby land ownership (instead of citizenship) discriminates between those who can access water and those who cannot, but it is also and deliberately naturalising the inequality ensuing from it. For instance, game farmers tend to oppose the physical redistribution of water on the basis of a natural limit, namely the hydrogeology of the Waterberg. Since aquifers are scattered unevenly across the plateau, they argue, it is only *natural* that some properties have a reliable water supply, while others do not. To transfer water from secluded farms into town would be practically and economically unfeasible and, above all, it would represent a blatant attack on private property rights.[55] In addition, foreign landowners seem to understand inequality as a *natural* feature of the South African landscape, so that the fact that (black) people in town have to queue at a tap at dawn in order to fill a bucket, whereas (white) people on a farm can enjoy water in abundance is simply perceived as 'the way things are'.[56] What is deeply problematic about these perspectives is not only their total lack of empathy for the living conditions of the majority of the local population of the Waterberg, but in particular their unravelling of a project of social production of space, whereby the place is valued and marketed as an unspoilt wilderness, whereas the presence of a growing mass of working poor and destitute people dependent upon social grants is perceived as highly *unnatural* and their water needs are disputed.

Conclusion

In this article I have questioned the mainstream view according to which game farms are saving water resources in the Waterberg (as opposed to irrigated crop farms) and have no relationship to the water problems in Vaalwater (caused by municipal inefficiency), while showing how green economy initiatives contribute to the reproduction and legitimation of inequality in access to water for local users. Indeed, some of those rural poor who suffer water shortages in the township of Leseding come from farms that were at some point converted into 'wilderness'. By moving – or being moved – into town, people have lost access to a secure, albeit limited, water supply. The point here is not whether municipal officials can deliver the limited resources at their disposal more efficiently. This is rather being used as a pretext to shift attention away from what is really at stake, namely how the citizenship rights of the rural poor and their role in the rural space are shaped by the process of conversion into private nature conservation. I have argued that nature conservation in the Waterberg is fundamentally built on the conservation of unequal power relations and this is the main reason why the redistribution of resources (land and water) is failing.

When framing the debate on water redistribution in terms of unequal power relations and social production of space – determining who is included and who is excluded from a place and its resources – it becomes clear that we also need to take the national level into account. The water question in the post-apartheid order has become fundamentally political and goes well beyond 'fixing' service delivery in small and under-resourced municipalities. It is the government and its national department, as custodians of the country's water resources, which have the legal and political means to produce change so that water access stops being a means to perpetuate discrimination among South African citizens. This calls for new research on both the politics of water redistribution and on the government's perception of what constitutes a just and equal society in contemporary South Africa.

Acknowledgements

This paper draws on a presentation given at the international conference Green Economy in the South, held at the University of Dodoma, Tanzania, in July 2014. I wish to thank the conference organisers and the editors of this special issue for creating a space of critical dialogue on such topic. I am grateful to Prof. Bram Büscher for his supervision and constant support. Thank you also to the two anonymous reviewers for their constructive and helpful comments, which permitted me to improve a first version of the article.

Notes

1. Waterberg Biosphere Reserve, *The Waterberg Meander*, 5.
2. According to UNEP's definition, a green economy is deemed to enhance economic growth while reducing environmental risks and ecological scarcities. UNEP, *Towards a Green Economy*, 16.
3. Hunter, *Pioneers of the Waterberg*; Taylor et al., *The Waterberg*; and Walker and Bothma, *The Soul of the Waterberg*.
4. See Cullis and van Koppen, *Applying the Gini Coefficient*; and Woodhouse, "Reforming Land and Water Rights."
5. A clear example of dualism is the 'two economies' rhetoric. For a political economy critique, see "Transcending Two Economies", the special issue of *Africanus* edited by Bond; and Marais, *South Africa*, 193–198.
6. Fairbairn et al., "Introduction."
7. Fairhead et al., "Green Grabbing"; and Mehta et al., "Introduction."
8. Büscher, "Letters of Gold"; and Fairhead et al., "Green Grabbing," 238.
9. Mehta et al., "Introduction."
10. O'Laughlin et al., "Introduction," 8; and Wegerif et al., *Still Searching for Security*, 41.
11. Du Toit, "Real Acts, Imagined Landscapes."
12. O'Laughlin et al., "Introduction," 4.
13. Brooks et al., "Creating a Commodified Wilderness"; and Snijders, "Wild Property." See also Spierenburg and Brooks, "Private Game Farming."
14. Snijders, "Wild Property."
15. Brandt and Spierenburg, "Game Fences in the Karoo."
16. Spierenburg and Brooks, "Private Game Farming."
17. Hall et al., "Farm Workers and Farm Dwellers," 53.
18. Langholz and Kerley, *Combining Conservation and Development*.
19. Snijders, "Wild Property"; and Spierenburg and Brooks, "Private Game Farming."
20. The conversion to game farming is only one of the many complex reasons explaining farm evictions during the post-apartheid era. Hall et al. interpret these in relation to overlapping and conflicting 'trajectories of change', namely agriculture restructuring and securing tenure for farm workers and dwellers. Hall et al., "Farm Workers and Farm Dwellers."
21. Li, "To Make Live"; and Spierenburg and Brooks, "Private Game Farming."
22. Snijders, "Wild Property," 506.
23. Personal communication, private nature reserve owner, June 28, 2014.
24. Vink and van Rooyen, *The Economic Performance*, 4.
25. Snijders, "Wild Property." The wildlife industry tends to make a distinction between conservation activities with no commercial purposes (nature reserves) and commercial activities based on the protection, but at the same time 'sustainable' use, of natural resources, such as eco-tourism, hunting and game breeding (game farms). However, this distinction is now blurring and in this article I employ the terms interchangeably.
26. Personal communication, local resident, May 9, 2014.
27. http://lapalala.com/the-new-development/, accessed March 4, 2015. Gianni Ravazzotti is the founder of Italtiles and was ranked among Africa's 40 Richest by *Forbes* in 2011. Peter Anderson is CEO of Anderson Wildlife Properties.
28. Waterberg District Municipality, http://www.waterberg.gov.za/docs/LED%20Brand.pdf, accessed March 2, 2015.
29. Personal communication, game farm manager, May 14, 2014.
30. At the time of writing, ZAR100 = US$8.5.
31. Snijders, "Wild Property," 512–513.
32. *Farmer's Weekly*, May 23, 2012, http://www.farmersweekly.co.za/article.aspx?id=22178&h=Talking-bull-with-Cyril-Ramaphosa.
33. Personal communication, game farm manager, June 11, 2014.
34. These are Terry McLintock, founder of Canon South Africa, and the insurance magnate Douw Steyn, respectively.
35. Personal communication, game farm manager, June 11, 2014.
36. On the website of Jembisa, now in the eco-tourism business, one can read that it all started with 'a family who wanted to provide their children with an African '"barefoot in the bush" childhood'. See http://www.jembisa.com/bushhome/the-bush-home/a-family-story.html, accessed March 3, 2015.
37. See Steinberg, *Midlands*.
38. In 2010 the Conservancy had 40 members in total, whereas in 2014 it had 70. Untitled draft document personally received from the Biosphere.
39. Personal communication, game farm owner, April 22, 2014.
40. A case in point is the Waterberg Welfare Society, established in 2000 by two local residents with the support of the Wilson Foundation, the charitable organisation of the American interior designer of luxury hotels, Trisha Wilson. In an interview Wilson commented: '22 years ago, I was awarded the Palace of

the Lost City project in South Africa's Sun City. That began my love affair with Africa [...] I ended up building a home in the Welgevonden game reserve. I became a member of the community, although I only visited there five weeks each year. You can't live in those communities and know those beautiful people and not get involved in fighting the poverty and disease.' *Interior Design*, June 18, 2013, http://www.interiordesign.net/articles/detail/34576-10-qs-with-trisha-wilson/.

41. Ibid.
42. See Ramutsindela et al., *Sponsoring Nature*.
43. National Archives of South Africa, TRB 2/1/651 124/0/65.
44. Black town residents were moved to the village of Steilloop, whereas farm workers were allowed to live on white farms and a few in a hostel in Vaalwater. National Archives of South Africa, HKN 1/1/19 HN9/15/3, TRB 217 4/0/65; and Rogerson and Letsoalo, "Resettlement and Under-development," 182.
45. According to the Census 2011, Vaalwater 'town' had a population of 3964 people and Leseding of 12,499. However, in 2012, the Vaalwater clinic registered a total population of 28,385 people. At the time of my research, water services were sourced from eight boreholes, which were bought or rented by the municipality along the years, with a total yield of 1.2 mega-litres per day.
46. MLM is both water service authority and service provider in Vaalwater. Although Modimolle town is 60 km distant to the south, Vaalwater was included in the municipality's borders following the new municipal demarcation process in 2000.
47. Personal communication, game farm owner, December 10, 2013.
48. These can be seen as the water equivalent of the constitutional property clause. However, following the phases of registration, validation, and verification, such uses are supposed to be granted (or denied) a water licence. Republic of South Africa, *National Water Act*, Chapter 1, Section 32.
49. Furthermore, the municipality finds it difficult to afford the price asked by farmers.
50. See Bond, "Theory and Practice."
51. Republic of South Africa, *National Water Act*.
52. Understandable in light of the fact that irrigation uses 60% of national water resources. DWA, *National Water Resource Strategy*, 9.
53. In this second case the reserve manager pointed out that they were very 'water conscious' and metered their water consumption. Also, they were registered with DWS, but did not pay water fees. Personal communication, game farm manager, November 5, 2013.
54. Republic of South Africa, *National Water Act*, Schedule 1.
55. One may note that infrastructure development faces hardly any limits on private reserves. Anyway, the point here is not to suggest that water needs to be physically redistributed but to show how the very idea of redistribution (of water and consequently land) is opposed. Personal communication, WNC general meeting, July 3, 2014.
56. Ibid.

Bibliography

Bond, Patrick, ed. "Transcending Two Economies – Renewed Debates in South African Political Economy." Special issue of the University of South Africa Development Studies journal *Africanus*, November 2007.

Bond, Patrick. "Theory and Practice in Challenging Extractive-oriented Infrastructure in South Africa." In *Sraffa and Althusser Reconsidered: Neoliberalism Advancing in South Africa, England, and Greece*, edited by Paul Zarembka, 97–132. Bingley: Emerald Group, 2014.

Brandt, Femke, and Marja Spierenburg. "Game Fences in the Karoo: Reconfiguring Spatial and Social Relations." *Journal of Contemporary African Studies* 32, no. 2 (2014): 220–237. doi:10.1080/02589001.2014.925300.

Brooks, Shirley, Marja Spierenburg, Lot van Brakel, Annemarie Kolk, and Khethabakhe B. Lukhozi. "Creating a Commodified Wilderness: Tourism, Private Game Farming, and 'Third Nature' Landscapes in Kwazulu-Natal." *Tijdschrift voor Economische en Sociale Geografie* 12, no. 3 (2011): 260–274. doi:10.1111/j.1467-9663.2011.00662.x.

Büscher, Bram. "Letters of Gold: Enabling Primitive Accumulation through Neoliberal Conservation." *Human Geography* 2, no. 3 (2009): 91–94.

Cullis, James, and Barbara van Koppen. *Applying the Gini Coefficient to Measure Inequality of Water Use in the Olifants River Water Management Area, South Africa*. Research Report 113. Colombo: International Water Management Institute, 2007.

Department of Water Affairs. *National Water Resource Strategy: Water for an Equitable and Sustainable Future*. Pretoria: DWA, 2013.

Du Toit, Andries. "Real Acts, Imagined Landscapes: Reflections on the Discourses of Land Reform in South Africa after 1994." *Journal of Agrarian Change* 13, no. 1 (2013): 16–22.

Fairbairn, Madeleine, Jonathan Fox, S. Ryan Isakson, Michael Levien, Nancy Peluso, Shahra Razavi, Ian Scoones, and K. Sivaramakrishnan. "Introduction: New Directions in Agrarian Political Economy." *Journal of Peasant Studies* 41, no. 5 (2014): 653–666. doi:10.1080/03066150.2014.953490.

Fairhead, James, Melissa Leach, and Ian Scoones. "Green Grabbing: A New Appropriation of Nature?" *Journal of Peasant Studies* 39, no. 2 (2012): 237–261. doi:10.1080/03066150.2012.671770.

Hall, Ruth, Poul Wisborg, Shirhami Shirinda, and Phillan Zamchiya. "Farm Workers and Farm Dwellers in Limpopo Province, South Africa." *Journal of Agrarian Change* 13, no. 1 (2013): 47–70.

Hunter, Elizabeth. *Pioneers of the Waterberg: A Photographic Journey*. Johannesburg: Camera Press CC, 2010.

Langholz, Jeffrey A., and Graham I. H. Kerley. *Combining Conservation and Development on Private Lands: An Assessment of Ecotourism-based Private Game Reserves in the Eastern Cape*. Report 56. Port Elizabeth: Centre for African Conservation Ecology, 2006.

Li, Tanya Murray. "To Make Live or Let Die? Rural Dispossession and the Protection of Surplus Populations." *Antipode* 41 (2009): 66–93. doi:10.1111/j.1467-8330.2009.00717.x.

Marais, Hein. *South Africa Pushed to the Limit: The Political Economy of Change*. London: Zed Books, 2011.

Mehta, Lyla, Gert Jan Veldwisch, and Jennifer Franco. "Introduction to the Special Issue: Water Grabbing? Focus on the (Re)Appropriation of Finite Water Resources." *Water Alternatives* 5, no. 2 (2012): 193–207.

O'Laughlin, Bridget, Henry Bernstein, Ben Cousins, and Pauline E. Peters. "Introduction: Agrarian Change, Rural Poverty and Land Reform in South Africa since 1994." *Journal of Agrarian Change* 13, no. 1 (2013): 1–15.

Ramutsidela, Maano, Marja Spierenburg, and Harry Wels. *Sponsoring Nature: Environmental Philanthropy for Conservation*. London: Earthscan, 2013.

Republic of South Africa. *National Water Act*. Act No. 36, 1998. Cape Town: Republic of South Africa, 1998. http://www.info.gov.za/view/DownloadFileAction?id=70693.

Rogerson, C. M., and E. M. Letsoalo. "Resettlement and Under-development in the Black 'Homelands' of South Africa." In *Population and Development Projects in Africa*, edited by John I Clarke, 176–193. Cambridge: Cambridge University Press, 1985.

Snijders, Dhoya. "Wild Property and its Boundaries – On Wildlife Policy and Rural Consequences in South Africa." *Journal of Peasant Studies* 39, no. 2 (2012): 503–520. doi:10.1080/03066150.2012.667406.

Spierenburg, Marja, and Shirley Brooks. "Private Game Farming and its Social Consequences in Post-apartheid South Africa: Contestations over Wildlife, Property and Agrarian Futures." *Journal of Contemporary African Studies* 32, no. 2 (2014): 151–172. doi:10.1080/09637494.2014.937164.

Steinberg, Jonny. *Midlands*. Johannesburg: Jonathan Ball, 2002.

Taylor, William, Gerald Hinde, and David Holt-Biddle. *The Waterberg: The Natural Splendours and the People*. Cape Town: Struik, 2003.

UNEP. *Towards a Green Economy: Pathways to Sustainable Development and Poverty Reduction*. Nairobi: United Nations Environment Programme.

Vink, Nick, and Johan van Rooyen. *The Economic Performance of Agriculture in South Africa since 1994: Implications for Food Security*. Working Paper 17. Midrand: Development Planning Division, Development Bank of Southern Africa, 2009.

Walker, Clive, and J. du P. Bothma. *The Soul of the Waterberg*. Houghton: Waterberg Publishers/African Sky Publishing, 2005.

Waterberg Biosphere Reserve. *The Waterberg Meander*. Vol. 1. Waterberg Biosphere Reserve, 2009. http://www.waterbergmeander.co.za/News_1_The+Waterberg+Meander+brochure.html.

Wegerif, Marc, Bev Russell, and Irma Grundling. *Still Searching for Security: The Reality of Farm Dweller Evictions in South Africa*. Polokwane/Johannesburg: Nkuzi Development Association/Social Surveys, 2005.

Woodhouse, Philip. "Reforming Land and Water Rights in South Africa." *Development and Change* 43, no. 4 (2012): 847–868. doi: 10.1111/j.1467-7660.2012.01784.x.

Extractive philanthropy: securing labour and land claim settlements in private nature reserves

Maano Ramutsindela

Department of Environmental & Geographical Science, University of Cape Town, Rondebosch, South Africa

At the centre of the conservation enterprise are the interactions of various actors who display a great deal of environmental ethic. Private landowners have embraced this ethic to protect their property rights and increase land value while contributing to the conservation of nature and to rural development. In this paper I draw examples from the lowveld in South Africa to argue that there is a seamless connection between philanthropy, labour and land claims in private nature reserves, and that post-apartheid conditions have enabled such a connection to emerge. Philanthropy allows private owners to structure and control labour, while directly or indirectly affecting the trajectory of land claims in the area.

Introduction

Environmental issues are a rallying point for groups, individuals and governments that are all concerned with the health of the planet and human survival thereon. This global concern has led to the development of global institutions and fora that seek to organise and support human responses to climate change, biodiversity protection, and so on. Indeed, the World Bank's Global Environment Facility, through which financial resources are channelled towards protecting the environment, has identified biodiversity conservation, global climate change, protection of international waterways and reduction of ozone-depleting chemicals as central to addressing the 'environmental crisis'. The private sector, one of the main culprits in this crisis as a result of its extractive activities, has come on board to fix environmental problems.[1] Whereas nature conservation in the form of national parks was historically seen as the remit of the state and its agencies, the private sector has shown growing interest in nature conservation.

The sector sees conservation areas as a niche market for capital accumulation, with companies in particular using it to develop their competitive advantage.[2] Two broad scholarly signatures on the connection between the private sector and the environmental crisis come from a pro-environmental lobby and from critical analyses. The pro-environmental lobby views the private sector as a source of funding crucial for solving environmental problems. This is evident in the call to place environmental tariffs on development projects in order to generate cash for conservation.[3] From the perspectives of critical scholarship, often but not always grounded in Marxism, participation of the private sector in environmental protection and conservation is driven by capitalist interests.[4]

There is a gap in critical scholarship in understanding how labour is involved in producing conservation goods and services. Sodikoff ascribes the analytic neglect of questions of labour in nature conservation to the imagination of conservation as an antithesis of production.[5] There is, however, a need to know what happens to labour when capitalism penetrates into conservation areas and infuses new systems of value. Moreover, we need to know how conservation labour is produced, organised and maintained as a result of such capital interest in conservation. It is clear that land use change from both subsistence and commercial agriculture entails a significant shift in the use of labour. For example, the workforce in commercial agriculture could become redundant when agricultural land is turned into, say, game ranches.[6] Similarly peasants' loss of land to conservation not only leads to loss of livelihood but also pushes peasants to sell their labour to economic sectors.[7] In this paper I focus on the seamless connection between philanthropy, labour and land claims in order to argue that private nature reserves use environmental philanthropy to achieve three interrelated objectives: to push back land claims, to give wealth-generating activities a human face, and to control a labour pool for purposes of upmarket ecotourism ventures. This form of extractive philanthropy is visible in private nature reserves in the lowveld (see below for a definition) near the Kruger National Park (KNP) in South Africa. The main goal of the paper is to contribute to research on conservation labour by drawing attention to the context under which such labour is produced and controlled. The paper draws on personal observations,[8] land claims court papers and research carried out in the lowveld in the past ten years.[9] Together, these sources are used to bring the question of labour into the analysis of nature conservation areas. First, however, it is important to clarify how environmental philanthropy becomes extractive before we dwell on the use of philanthropy to control labour and influence land claims in private nature reserves in South Africa.

Philanthropy, labour and land claims

At face value the links between philanthropy, labour and land claims seem odd. This is understandable, because philanthropy is associated with the generosity of the human spirit. Indeed, philanthropy and its equivalent terms (charity, benevolence, giving, donation, voluntary sector, third sector, independent sector) is generally understood as a 'voluntary action for the public good'. It underpins human interventions in various fields ranging from the art to educational institutions.[10] Given this general understanding of philanthropy, how do we connect it to issues of labour and land claims, and how does such connection help us understand conservation labour as a process by which labour serves the conservation enterprise? The few attempts at explaining conservation labour have focused on the division of labour between highly paid (often foreign) experts and low-paid (mostly local) workers, the transfer of manual labour from agriculture to conservation and the consequent tensions, and the recycling of colonial labour practices in nature conservation.[11] While Sodikoff's notion of the low-wage conservationist is helpful for interrogating questions of labour in the context of conservation, I suggest that analyses of conservation labour can be advanced by grasping ways in which labour is structured and controlled. I demonstrate this by placing labour control within the context of philanthropy in private nature reserves. These reserves are involved in business ventures where both conservation and profit-making intersect. Here labour is not only required for the business of ecotourism but is entangled with philanthropy and the political process of land restitution.

Such entanglement can best be understood through a conceptual triangulation of philanthropy, labour and land claims. The main requirements for such triangulation are that philanthropic activities should be understood as reflecting the interests and biases of philanthropists,[12] and also as promoting values of social order and economic development favourable to those holding political and/or economic power.[13] Philanthropy has, for instance, been used by powerful states as an instrument of foreign policy, an avenue to influence the behaviour of developing countries, and to promote capitalism. Major US foundations, such as Ford, Rockefeller and Carnegie, played a significant role in promoting US foreign policy interests in Africa in the postwar era.[14] Schramm notes that philanthropy 'is intended to apply wealth in ways that preserve democratic pluralism and a free-market economy, thereby promoting the "release of human possibilities" as intended by foundation donors'.[15] In the same vein Sawaya sees the current version of philanthropy as a flashpoint for debates about liberal capitalism.[16] My concern here is with how the intersection between philanthropy and capitalism, which Bishop calls philanthrocapitalism,[17] manifests in nature conservation projects and how it also shapes labour.

The 'environmental crisis' has magnified the link between philanthropy and nature conservation. In turn the link gave rise to the emergence of environmental philanthropy, which is understood as a subfield of philanthropy that 'encompasses resources that individuals, communities, the business sector and foundations commit to the preservation and conservation of nature and the promotion of activities related to nature conservation and the general health of the planet'.[18] Philanthropists use their financial resources to purchase land for conservation purposes, to establish and protect parks, conservancies and nature reserves, and to fundraise for conservation trusts and various kinds of conservation projects.[19]

As with all other forms of philanthropy, environmental philanthropy is not only selective according to the environmental cause that the giver supports, but can also result in unintended negative consequences. For example, much of the land on which national parks in the former colonies were established belonged to or was used by locals whose land rights were violated by conservation organisations and state agencies.[20] This practice continued under the watchful eye of post-independence governments,[21] and in projects such as peace parks, which claim to care about people holding land under various forms of tenure.[22] Conservation projects such as peace parks that straddle the borders of two or more countries receive massive funding from the private sector and philanthropists,[23] yet their negative consequences (eg forced removals or compulsory excising of communal land for purposes of enlarging conservation areas[24] do not seem to be of concern to governments or the philanthropists involved. With nature conservation as the main goal, the harms of environmental philanthropy appear as either unintended consequences or the cost necessary for a global cause, ie the health of the planet for the survival of human and non-human species.

The participation of the private sector in the conservation enterprise brings questions of environmental philanthropy and labour to the forefront, not least because the private sector is profit-driven irrespective of whether it pursues a conservation cause or not. A more compelling reason why questions around labour are critical for assessing environmental philanthropy is that private nature reserves pursue both conservation and business goals simultaneously, while also requiring a labour pool that supports the business plan. This is particularly so in private nature reserves involved in high-end ecotourism ventures.

Philanthropy is extractive both materially and discursively, and its extractive nature is laid bare when altruism is overtaken by political and economic interests. Materially business exploits labour for profit maximisation but then use philanthropy to address the social ills created by the business enterprise. This makes philanthropy part of the process of labour exploitation. Thus wealth generated, say, through cheap labour is channelled through philanthropy to cushion the consequences of labour exploitation in the same way that foreign aid was used to soften the blows of structural adjustment in African countries.[25] As we shall see below, philanthropy may be used to structure and control labour and to influence the outcomes of land claims through discourses of development. Such discourses focus on, and channel philanthropic resources towards poverty alleviation in part to disengage local communities from land-reform processes.

Experiences in South Africa's lowveld

In this paper the lowveld refers to the low-lying region (150–600 meters above sea level) in the provinces of Limpopo and Mpumalanga in the vicinity of the Kruger National Park. Since its proclamation in 1898 the KNP has not only been one of Africa's most well-known parks, but also the economic backbone of private nature reserves whose identity and market are tied to the KNP. For example, on its website, Sabi Sabi Game Reserve states that:

> Sabi Sabi Private Game Reserve is an unspoilt part of Africa ecologically and *geographically integrated with the adjacent Kruger National Park*. It is situated within one of South Africa's oldest and largest proclaimed reserves – the renowned Sabi Sand Wildtuin (Author's emphasis).[26]

Another private nature reserve, Ngala, presents itself as part of the famous KNP:

The Kruger National Park has nearly two million hectares (almost five million acres) of unfenced African wilderness, in which more mammal species roam free than in any other game reserve. A leader in environmental management, Kruger offers visitors fantastic Big Five sightings, as well as viewing of endangered game such as the African wild dog and birdwatching of over 507 bird species. *Within the Kruger lies &Beyond Ngala Private Game Reserve*, the first private game reserve to be incorporated into this famous park (Author's emphasis).[27]

Private nature reserves adjacent to the KNP took a major turn in the 1990s when the owners of these reserves established a tourism industry to take advantage of the end of apartheid rule in the country to promote ecotourism. They were of course also sensitive to the political ramifications of the legacies of apartheid for their business ventures. Led by Londolozi game reserve, owned by the Varty brothers,[28] private nature reserves sought to bring together the objectives of land rehabilitation, business and community development.[29] These objectives are at the core of the discursive form of extractive philanthropy mentioned above.

The development of these nature reserves and the business strategies they pursued from the early 1990s onwards cannot be isolated from apartheid and the birth of a democratic South Africa. For example, the western part of KNP, where private nature reserves are concentrated, was highly degraded because of the apartheid government's neglect of this area, and also as a result of a high population concentration in the ecologically fragile former Gazankulu bantustan.[30] The area of Gazankulu adjacent to the KNP was and still is highly impoverished. In addition to this ecology of apartheid, some private nature reserves, such as Mala Mala, were established on land from which black people were forcibly removed by the apartheid government. Below I discuss how land claims in these reserves have been handled.

The Londos strategy[31]

The evolution of the ecotourism strategy known as the 'Londos strategy' trace back to the land-use change from hunting to ecotourism. This shift is ascribed to 'Boyd Varty, son of the original owner, [who] almost on his death bed, made the statement that hunting must stop and the land must pay for itself'.[32] The strategy should also be understood as a response to changing political conditions, accompanied by threats to commercial farms,[33] and business opportunities opened by the dawn of democracy. Accordingly the business model of private nature reserves was sensitive to both the ecology of the area and to socio-political realities.

In the lowveld, privately owned nature reserves such as Londolozi embarked on the rehabilitation of ecological systems as the basis for the business of ecotourism.[34] The reserve owners envisaged a type of ecotourism that supports the conservation of the environment by minimising the ecological impact of tourists. It is for this reason that these private nature reserves target few but wealthy tourists through a pricing system that makes the lodges therein unaffordable to the majority of the population in the country. At face value, high prices suggest value for money, a unique experience offered by the reserves and concern with preserving the ecology as a business product.[35] Such high prices, however, also mean two other important things. First, they render the lodges inside these private nature reserves exclusive spaces on financial rather than, say, racial grounds. Second, the prices target wealthy individuals, some of who contribute to charitable activities carried out by private nature reserves. A closer inspection of the model reveals how this works in practice.

The Varty brothers promoted an ecotourism business that contributes to social and economic development of communities neighbouring their properties. This approach, which Koch calls the 'Londos strategy', represents a business strategy that combines elements of social corporate social responsibility, philanthropy and black economic empowerment.[36] The Londos strategy was carried out through the Conservation Corporation Africa (CCA), founded in 1992 by 'a group of individual shareholders'.[37] The CCA is considered 'the largest private company involved in wildlife tourism in South Africa'.[38] It sought to capture the expected increase in tourism demand in post-apartheid South Africa. By 1995 the company owned 26 accommodation-type properties in Botswana, Kenya, Namibia, Tanzania, South Africa and Zanzibar.[39] The founding principles of the company encapsulated three areas of care: care of the land, care of the wild and care of the people.

The notion of caring for the land raises the question of whether private property is most appropriate for protecting the land and the biodiversity thereon. The debate on the implications of various forms of land ownership (tenure regimes) for biodiversity protection and management is still going on, with some believing that private land tenure and the recognition of property rights hold much promise for the future of biodiversity.[40] The question that most analysts of ecotourism neglect is how private nature reserves involved in this type of tourism affect land issues. This question is pertinent to South Africa because of the history of land dispossession.

Pushing back land reform

A proper evaluation of ecotourism in private nature reserves and on state land such as national parks should include land rights issues. I argue that the business model described above uses philanthropy to push back the quest for land restoration by 'doing good work' for existing and potential land claimants. This renders philanthropy extractive in that charity is used to advance the interests of private landowners at the expense of victims of apartheid-era forced removals. It was clear to white property owners in South Africa in the early 1990s that land redistribution would be unavoidable in post-apartheid South Africa because of the highly racially skewed land ownership patterns in the country.[41] The majority of owners of private nature reserves in the lowveld view land claims negatively, as evident in a statement issued by Londolozi:

> *We and other owners* in the reserve continue to dispute the validity of land claims in the Sabi Sand. We have advised the community that any land claims against Londolozi will have to be fully tested in court. For obvious reasons, however, we are willing to assist the community should they become our new neighbours (Author's emphasis).[42]

Fourteen land claims were lodged on Sabi Sabi Wildtuin before the 1998 deadline for the lodgement of land claims in the country.[43] These claims include that of Mhlanganisweni community on the farms of 'Sparta and Marthly on which

Londolozi operates its game viewing activities'.[44] As in most land claims against private property, the owners of Londolozi opposed Mhlanganisweni land claims. They sought to invalidate the claim after it was gazetted in 2002 through its consultant,[45] who reasoned that the claimants were made up of people coming from five different villages and four tribal authorities and therefore did not constitute a homogeneous community eligible to claim the land.[46] It was also argued that the claimants lived on the land as farm labourers or tenants and therefore did not have the status of landowners, as if such a status existed in the Gazankulu bantustan under apartheid. Were Mhlanganisweni land claim to succeed, it would lose Londolozi's discretionary contribution to rural development that has already been built into the Londos strategy. Opponents of the claim argue that it is highly unlikely that the community will succeed in operating such an upmarket ecotourism business.

Mala Mala broke ranks with other landowners in the Sabi Sabi Wildtuin by acquiescing to the land claim.[47] In doing so, however, the owners sought to 'give with one hand and take with another'.[48] I argue that pricing in private nature reserves is not only meant to exclude ordinary citizens from the reserves but should also be understood as a mechanism for securing land even in the face of land reform. The land claim against Mala Mala private game reserve in the lowveld is a case in point. This claim comprised of multiple claimants who lodged land claims on various farms comprising Mala Mala Game Reserve in 1996.[49] The land claims were validated but proved difficult to settle, mainly because of the price of the properties involved. After subjecting pieces of land in Mala Mala to property valuation, the provincial government of Mpumalanga offered to buy land on behalf of the claimants on a price based on land value per hectare and value of improvements made (Table 1).

The owners of Mala Mala rejected the offer on the grounds that the price offered by the government was lower than the actual value of the properties. This created a deadlock that could have been resolved by land expropriation by government.[50] Amid mounting criticism of the lack of political commitment to land reform by the ruling African National Congress, and the forthcoming fifth national elections held in 2014, the state offered to buy Mala Mala Game Reserve at the cost of R1.1 billion.[51]

Table 1.   Mala Mala Property Values in government's initial offer (After Shabangu 2014).

Farm	Valuer	Land Value per hectar	Value of fixed Improvements	Total
Remaining Extent and Portion 1 of Eyerfield 343 KU	Dijalo Property Valutions	R 65 000/ha		R194, 000,000
Portion 7 of Toulon 383 KU	Dijalo Property Valutions	R 65 000/ha		R 27, 000, 000
Mala Mala 341 KU	Dijalo Property Valutions	R 65 000/ha	R 33, 528,850	R 153,000,000
Mala Mala 359 KU	Dijalo Property Valutions	R 65 000/ha	R 9,354,642	R 92 000,000
RE Charleston 378 and Portion 1 Charleston 375 KU	Bristow, Phenyane and Associates	R 66,627 /ha	R 1,253,000	R 241,253,000
Flockfield 414 and 361 KU	Fincon (J A van Rensburg)	R 30,000 /ha	R 22,033,000	R 105,000,000
			R 66,169,492	R 812,253,000

Following the purchase of the game reserve Mala Mala PTY (Ltd) proposed to form a partnership with Mhlanganisweni community in which the community would keep Mala Mala as a brand already well known internationally, while the company would contribute its movable assets falling outside the R1.1 billion the government had paid. It further proposed share schemes that will see the community acquiring an additional 10% of shares over a 50-year period. In the interim the community has signed a lease agreement with Mala Mala PTY in which the community will earn R700 000 per month until a long-term agreement intended to protect the investment made by government in purchasing the land is finalised.[52]

I refer to land restitution above to emphasise that most owners of private nature reserves in the lowveld are fiercely opposed to land restitution. They instead prefer other measures, such as rural development programmes, which keep ownership of their land intact. This strategy reflects broader debates on whether land restitution should be prioritised in the face of major failures in land reform projects.[53] There is a view that giving rural dwellers opportunities such as better education and jobs might be more beneficial than pursuing the political agenda of land reform.[54] In the context of private nature reserves this view is promoted not only to push back land restitution but also as an avenue for philanthropic activities in rural areas adjacent thereto. In turn, these activities are instrumental in forming relationships between the reserves and neighbouring communities to ensure the successful operation of business in nature reserves.

Philanthropy for development and labour control

> We believe that only through achieving socio-economic development and growth will our eco-tourism industry be ensured of legitimacy in the new order of South Africa. Our approach is, we believe, the only means of ensuring that our wilderness areas remain economically viable.[55]

Philanthropy formed a strong element of the Londos strategy right from the creation of the CCA in 1992. In order to care for people and to turn them into 'friendly neighbours', the CCA established a Rural Investment Fund 'to raise capital for, and promote, major infrastructural projects in the [economically] depressed areas around [private nature reserves]'.[56] Most of the private and bilateral donor funds channelled to community development projects came from the USA, Canada and Europe.[57] A philanthropic Africa Foundation was formed in 1992 'to uplift, up-skill and empower rural communities living primarily adjacent to conservation areas' in order to protect Africa's heritage.[58] The foundation sees its role as facilitating the fulfilment of needs identified by communities, communicating those needs to potential donors, allocating and managing donor funds prudently, working with community leaders and project champions, accounting and reporting to donors, and evaluating the short- and long-term impacts of the foundation's projects.[59] It is supported by other branches of the foundation, namely Africa Foundation (USA) and Africa Foundation (UK), and has South Africa's world renowned Emeritus Archbishop Desmond Tutu and his wife, Leah, as its patrons.

As in most NGOs that provide a service to casualties suffered by the poor as a result of business operations, the CCA's approach to local development is a response to the fear that high unemployment in villages surrounding game reserves will result in poaching of game. The solution has been to extend a hand to local communities so that they can police themselves.[60] As part of its philanthropic initiatives Londolozi contributes to the improvement of education in surrounding areas. For example, it has built 18 classrooms, held debates between secondary schools and raised funds for scholarships and the development of computer training facilities.[61] Of significance to the theme of this paper is how labour is secured through philanthropy.

When photographic tourism took off in Londolozi in the 1970s, local people working for Londolozi were allowed to live inside the game reserve, although they 'all had permanent homesteads in the communities on the western and northern borders of the SSW'.[62] From the 1980s the number of workers living inside the reserve grew as more labour was needed for the growing business. Koelble estimates that 135 employees of Londolozi live on site.[63] Hendry notes that 'most staff lived on site at Londolozi and spent roughly 75 percent of their time on the reserve and the rest at home with their families in the villages where they had their homes'.[64] The result of this arrangement was the emergence of a migrant labour system in Londolozi and a two-pronged approach to community development through philanthropy.

Whereas the initial focus of community development was on alleviating poverty in surrounding communities, the increase in demand for labour in Londolozi led to a redefinition of both the local community and development. The 'local' became people coming from surrounding villages but living inside the nature reserve, for whom philanthropy was needed for skills development and social well-being. Community development in Londolozi thus moved away from trying to improve the socioeconomic situation of nearby villages to creating its own village inside the reserve.[65]

A number of reasons are used to argue for community development inside Londolozi. For example, it is held that community development projects such as infrastructural development in communities adjacent to private game reserves have been inefficient and sometimes wasteful.[66] Investment in community projects inside the reserve allows for better monitoring of progress and accountability. It is also believed that, by focusing on people living inside the reserve, the business skills required by Londolozi and the professional development of an individual staff member can be harmonised to ensure the efficiency and profitability of the business operation. Furthermore, Londolozi is able to showcase community development projects to international guests with strong values of responsible tourism. Thus community development projects inside the reserve also serve as a marketing tool for Londolozi. From the perspective of capitalism housing workers inside the reserve can be seen as a cost-cutting measure to ensure maximum profitability. The cost of transporting workers is reduced, while the availability of labour is guaranteed. There is some parallel here with practices on South Africa's white farms, where farmers have a tendency to keep farm workers on site to achieve similar objectives.

Conclusion

In this paper I have explored the seamless connection between philanthropy, labour and land claims in private nature reserves in South Africa to contribute to the agenda for research on conservation labour. Identifying Sodikoff's low-wage conservationist is a useful starting point but there is a need to expand questions of labour in nature conservation to understand how labour is produced, structured and controlled. As I have demonstrated, philanthropy facilitates the control of labour in private nature reserves and is also used to influence the outcomes of land restitution. In doing so, it mediates land–labour relationships. The use of philanthropy to achieve these goals goes unnoticed mainly because philanthropy is generally associated with the public good. It nevertheless point to the negative consequences (harms) of philanthropic activities that require scholarly attention. Environmental philanthropy is involved in the control of conservation labour, although this may not be the intention of the philanthropists involved.

The owners of private nature reserves such as Londolozi initiated developmental projects in neighbouring communities in order to reconstruct neighbourhoods mainly for the economic benefit of the reserve. The use of philanthropy to achieve such a goal renders philanthropy extractive in that the founding values of altruism are sacrificed and discourses of development mobilised and financed through philanthropy for capital accumulation. The Londos strategy and the philanthropy associated with it has nevertheless gone some way in assisting rural residents whose conditions of life are shaped by the legacies of apartheid and the post-apartheid government's inability to overcome them.

In the process of helping the poor, philanthropy in nature reserves in South Africa's lowveld speaks to three logics, namely caring for nature, socioeconomic development of the rural poor and calculated responses to land claims. It is used as a response to land restitution to ensure the integrity of private nature reserves and the success of ecotourism business operations. Land restitution is highly emotive and deeply political, and is largely seen by private land owners as a threat to property. I conclude that philanthropy in the lowveld is self-serving in two main ways. First, it is used to finance community development projects while also 'taming' land claimants. Second, shifting the focus away from communities outside the reserve towards those living inside has material consequences. Communities who press ahead with land claims stand to lose philanthropic support initiated by landowners. Residents living inside the nature reserve are not simply beneficiaries of philanthropy but constitute a labour pool that grows with the growth in the ecotourism business. The settlement inside the reserve makes it easier to control and monitor labour. Conceptually the growth in ecotourism calls for attention to various ways in which conservation labour is produced and maintained.

Notes

1. This move by the private sector has been the subject of much scholarly discussion. See Castree, "Neoliberalising Nature"; and Büscher et al., "Towards a Synthesized Critique."
2. Brockington and Duffy, "Capitalism and Conservation"; and Porter and Kramer, "The Competitive Advantage."
3. Snyder, "Editorial."
4. Castree, "Neoliberalism and the Biophysical Environment"; and Büscher et al., "Towards a Synthesized Critique."
5. Sodikoff, "The Low-wage Conservationists."
6. Brandt and Spierenburg, "Game Fences in the Karoo."
7. The clearest example of this was the dispossession of land under apartheid to force blacks to work in the mines.
8. I was intrigued by a 'village' inside Londolozi on my first visit to the reserve in 2008. Subsequently I researched environmental philanthropy and land reform in separate projects.
9. I summarise findings from research carried out in the lowveld to develop the argument of this paper. I guided some of the research referred to in the paper.
10. Ilchman et al, *Philanthropy*, x. See also Adam, "Introduction."
11. Sodikoff, "Forced and Forest Labor"; and West, *Conservation is our Government Now*.
12. Holmes, "Biodiversity for Billionaires."
13. Damon, *The Moral Advantage*.
14. Berman, *The Ideology of Philanthropy*; and Clotfelter and Ehrlich, "The World we must Build."
15. Schramm, "Law outside the Market," 359–360.
16. Sawaya, "Capitalism and Philanthropy," 202.
17. Bishop, "What is Philanthrocapitalism?"
18. Ramutsindela et al., Sponsoring Nature. Holmes prefers the term conservation philanthropy. Holmes, "Biodiversity for Billionaires."
19. Delfin and Tang, "Elitism, Pluralism"; and Butler et al., *Wildlands Philanthropy*.
20. Dowie, *Conservation Refugees*.
21. Hutton et al., "Back to the Barriers?"
22. Ramutsindela, Transfrontier Conservation.
23. Ibid.
24. Milgroom and Spierenburg, "Induced Volition."
25. Bornstein, *The Spirit of Development*.
26. http://www.sabisabi.com/information/location, accessed March 9, 2015.
27. http://www.andbeyond.com/south-africa/places-to-go/kruger-national-park-surrounds.htm, accessed March 9, 2015.
28. The brothers John and Dave Varty are the third generation of the family that owned Sparta farm, from which Londolozi originated. Londolozi forms part of Sabi Sand Wildtuin (SSW), a nature reserve consisting of a consortium of exclusive game lodges on privately owned land abutting the KNP. The farm on which Londolozi is built, Sparta, has been in the hands of the Varty family since 1926. The farm was originally used for hunting and became a premier photographic safari lodge in the 1970s. Varty, *Full Circle*.
29. All these objectives were pursued at the height of liberation struggles in South Africa in the late 1980s and at the dawn of democracy in the country in the early 1990s.
30. Bantustans were areas designated for occupation and use by the African population during apartheid. Most former bantustans were highly degraded, not so much because people did not know how to look after the land but mainly because these political entities were established on marginal lands on which high populations of mostly impoverished people were to eke out a living. Ramphele and McDowell, *Restoring the Land*; Hoffman and Ashwell, *Nature Divided*; and Hebinck and Lent, *Livelihoods and Landscapes*.
31. Londolozi made the first major breakthrough in developing ecotourism from land previously used for agriculture and hunting.
32. Hendry, "Nature Conservation."
33. In the early 1990s South Africa was gripped by fear of a potential civil war and there were also anxieties about how the political transition would play out. The white population feared that black people might take revenge by, for example, taking away white farms from which blacks had been forcibly removed during apartheid. Their concern was heightened by sporadic land invasions that took place during the political transition from apartheid rule.
34. Koelble claims that 'Londolozi pioneered the concept of ecotourism and popularized the approach through its operations'. Koelble, "Ecology, Economy and Empowerment," 7.
35. A degraded environment is not good for ecotourism.
36. Koch, "Ecotourism and Rural Reconstruction," 227.
37. Koelble, "Ecology, Economy and Empowerment," 9.
38. Wells, "The Social Role of Protected Areas," 326.

39. Groch et al., "The Grassroots Londolozi Model."
40. Brubaker, *Property Rights*.
41. This was the time the Londos strategy was devised.
42. http://www.tourismupdate.co.za/home/detail?articleid=24958&article=letter-to-the-editor-londolozis-re sponse-on-land-claim-issue, accessed March 14, 2015. Levin and Weiner, *No More Tears* provide a detailed analysis of land claims in the area.
43. The date for lodging land claims was reopened in 2014 through the passage of the Land Restitution Amendment Act.
44. Hendry, "Nature Conservation," 71.
45. The consultant J. B. Hartman investigated the merits of land claims on seven farms in the Sabi Sabi Wlidtuin and produced the report in 2003.
46. Hendry, "Nature Conservation."
47. "Letter to the Editor: Londolozi's Response on Land Claim Issue," April 12, 2012. http://www.tourismupdate. co.za/home/detail?articleid=24958&article=letter-to-the-editor-londolozis-response-on-land-claim-issue.
48. As will be clear below, while the owners of Mala Mala agreed to sell their property to government at an exorbitant price, they structured the land deal in such a way that they could continue to operate their business on land that had been restored. Thus, they gave away the land but retained control of the business. The owners of Mala Mala, Michael and Norma Rattray, assured their friends and clients that the settlement of the land claim did not change their control over the business operation of the reserve. "Statement released by Mala Mala Game Reserve," January 17, 2014. I consider this a subtle method of resisting land reform in South Africa. This method is not exclusive to private nature reserves but also permeates partnerships formed over other land restitution projects.
49. Land Claims Court of South Africa, "Case number LCC156/2009."
50. According to Section 25 of the Constitution of the Republic of South Africa, the government can appropriate land in the interest of the public.
51. There is an anonymous rumour that a property close to Mala Mala was sold at a higher than market value during negotiations between government and Mala Mala PTY in order to inflate the price that the government finally paid.
52. Shabangu, "The Neo-liberalization of Nature."
53. Cousins and Walker, *Land Divided Land Restored*.
54. Cherryl Walker made remarks along these lines at the launch of the book, *Land Divided Land Restored* at the Book Club in Cape Town on February 25, 2015.
55. Conservation Corporation news, cited in Koch, "Ecotourism and Rural Reconstruction." The strategy for ensuring the legitimacy of these private nature reserves included getting former President Nelson Mandela to visit Londolozi, where he is said to have praised it as a progressive game reserve in which 'people of all races [live] in harmony amidst the beauty that mother nature offers'. Varty and Buchanan, *I Speak of Africa*.
56. Koch, "Ecotourism and Rural Reconstruction," 227.
57. Burns and Barrie, "Race, Space and 'Our Own Piece of Africa'."
58. www.africafoundation.org.za/the-heart/who-we-are, accessed March 10, 2015. According to its website, Africa Foundation is currently registered as a South African Trust (registration number: IT2542/1993); a Public Benefit Organisation (PBO) (registration number: 930002115); and a not-for-profit organisation (registration number: 004-145).
59. Ibid.
60. See Burns and Barrie, "Race, Space and 'Our Own Piece of Africa'."
61. Koelble, "Ecology, Economy and Empowerment."
62. Personal communication between Varty and James Hendry, May 10, 2008.
63. Koelble, "Ecology, Economy and Empowerment."
64. Hendry, "Nature Conservation," 44.
65. It is said that Londolozi adopted this approach in 2008 following the 2005 survey that was used to collect the views and developmental aspirations of its staff.
66. For example, 'there were new classrooms strewn with litter and broken furniture, defunct vegetable gardens, disused boreholes, rooms full of unused computers at schools, boxes of unopened books, an unused library and various other projects in disrepair'. Hendry, "Nature Conservation," 54.

Bibliography

Adam, T. "Introduction." In *Philanthropy, Patronage and Civil Society: Experiences from Germany, Great Britain, and North America*, edited by T. Adam, 1–12. Bloomington: Indiana University Press, 2004.
Berman, E. H. *The Ideology of Philanthropy: The Influence of the Carnegie, Ford and Rockefeller Foundations on American Foreign Policy*. Albany: State University of New York, 1983.
Bishop, M. "What is Philanthrocapitalism?" *Alliance* (Mar. 2007): 30.
Bornstein, E. *The Spirit of Development*. London: Routledge, 2003.

Brandt, F., and M. Spierenburg. "Game Fences in the Karoo: Reconfiguring Spatial and Social Relations." *Journal of Contemporary African Studies* 32, no. 2 (2014): 220–237.

Brockington, D., and R. Duffy. "Capitalism and Conservation: The Production and Reproduction of Biodiversity Conservation." *Antipode* 42, no. 3 (2010): 469–484.

Brubaker, E. *Property Rights in Defence of Nature.* London: Earthscan, 1995.

Burns, P. M., and S. Barrie. "Race, Space and 'Our Own Piece of Africa': Doing Good in Luphisi Village?" *Journal of Sustainable Tourism* 13, no. 5 (2005): 468–485.

Büscher, B., S. Sullivan, K. Neves, J. Igoe, and D. Brockington. "Towards a Synthesized Critique of Neoliberal Biodiversity Conservation." *Capitalism Nature Socialism* 23, no. 2 (2012): 4–30.

Butler, T., A. Vizcaíno, and T. Brokaw. *Wildlands Philanthropy: The Great American Tradition.* San Rafael, CA: Earth Aware.

Castree, N. "Neoliberalism and the Biophysical Environment 1: What 'Neoliberalism' is, and what Difference Nature makes to It." *Geography Compass* 4, no. 12 (2010): 1725–1733.

Castree, N. "Neoliberalising Nature: Processes, Effects and Evaluations." *Environment and Planning A* 40, no. 1 (2008): 153–173.

Clotfelter, C. T., and T. Ehrlich. "The World we must Build." In *Philanthropy and the Nonprofit Sector in a Changing America,* edited by C. T. Clotfelter and T. Ehrlich, 499–516. Bloomington: Indiana University Press, 1999.

Cousins, B., and C. Walker. *Land Divided Land Restored: Land Reform in South Africa for the 21st Century.* Aukland Park: Jacana, 2015.

Damon, W. *The Moral Advantage: How to Succeed in Business by Doing the Right Thing.* San Francisco, CA: Berrett-Koehler, 2004.

Delfin, F. G., and S. Tang. "Elitism, Pluralism, or Resource Dependency: Patterns of Environmental Philanthropy among Private Foundations in California." *Environment and Planning A* 39, no. 9 (2007): 2167–2186.

Dowie, M. *Conservation Refugees: The Hundred-year Conflict between Global Conservation and Native Peoples.* Cambridge, MA: MIT Press, 2009.

Groch, K., K. E. Gerdes, E. A. Segal, and M. Groch. "The Grassroots Londolozi Model of African Development: Social Empathy in Action." *Journal of Community Practice* 20, nos. 1–2 (2012): 154–177.

Hebinck, P. G. M., and P. C. Lent, eds. *Livelihoods and Landscapes: The People of Guquka and Koloni and their Resources.* Brill: Leiden, 2007.

Hendry, J. R. A. "Nature Conservation in Changing Socio-political Conditions at Londolozi Private Game Reserve." Masters diss., University of Cape Town, 2008.

Hoffman, M. T., and A. Ashwell. *Nature Divided: Land Degradation in South Africa.* Cape Town: University of Cape Town Press.

Holmes, G. "Biodiversity for Billionaires: Capitalism, Conservation and the Role of Philanthropy in Saving/ Selling Nature." *Development and Change* 43, no. 1 (2012): 185–203.

Hutton, J., W. M. Adams, and J. Murombedzi. "Back to the Barriers? Changing Narratives in Biodiversity Conservation." *Forum for Development Studies* 2 (2005): 341–370.

Ilchman, W. F., S. N. Katz, and E. L. Queen. *Philanthropy in the World's Traditions.* Bloomington: Indiana University Press, 1998.

Koelble, T. A. "Ecology, Economy and Empowerment: Ecotourism and the Game Lodge Industry in South Africa." *Business and Politics* 13, no. 1 (2011). doi: 10.2202/1469-3569.1333.

Koch, E. "Ecotourism and Rural Reconstruction in South Africa: Reality or Rhetoric?" In *Social Change & Conservation,* edited by K. Ghimire and M. P. Pimbert, 214–238. London: Earthscan, 1997.

Levin, R., and D. Weiner eds. *No more Tears: Struggles for Land in Mpumalanga, South Africa.* Trenton, NJ: Africa World Press, 1997.

Milgroom, J., and M. Spierenburg. "Induced Volition: Resettlement from the Limpopo National Park, Mozambique." *Journal of Contemporary African Studies* 26, no. 4 (2008): 435–448.

Porter, M. E., and M. R. Kramer. "The Competitive Advantage of Corporate Philanthropy." *Harvard Business Review* (Dec. 2002): 56–68.

Ramphele, M., and C. McDowell. *Restoring the Land: Environment and Change in Post-apartheid South Africa.* London: Panos, 1991.

Ramutsindela, M. *Transfrontier Conservation in Africa: At the Confluence of Capital, Politics and Nature.* Wallingford, CT: Cabi, 2007.

Ramutsindela, M., M. Spierenburg, and H. Wels. *Sponsoring Nature: Environmental Philanthropy for Conservation.* London: Earthscan, 2011.

Sawaya, F. "Capitalism and Philanthropy in the (New) Gilded Age." *American Quarterly* 60, no. 1 (2008): 201–213.

Schramm, C. J. "Law outside the Market: The Social Utility of the Private Foundation." *Harvard Journal of Law and Public Policy* 30, no. 1 (2006): 355–415.

Shabangu, M. "The Neo-liberalization of Nature: Contextualizing the Resolution of Land Claims in the Kruger National Park." Masters diss., University of Cape Town, 2014.

Snyder, B. F. "Editorial: Solving Conservation's Money Problems." *Conservation Biology* 29, no. 1 (2015): 1–2.

Sodikoff, G. "The Low-wage Conservationists: Biodiversity and Perversities of Value in Madagascar." *American Anthropologist* 111, no. 4 (2009): 443–455.

Sodikoff, G. "Forced and Forest Labor Regimes in Colonial Madagascar, 1926–1936." *Ethnohistory* 52, no. 2 (2005): 407–435.

Varty, D. *Full Circle*. Johannesburg: Penguin, 2008.

Varty, S., and Buchanan, M. *I Speak of Africa*. Johannesburg: Londolozi Publishing, 1997.

Wells, M. P. "The Social Role of Protected Areas in the New South Africa." *Environmental Conservation* 23, no. 4 (1996): 322–331.

West, P. *Conservation is our Government Now: The Politics of Ecology in Papua New Guinea*. Durham, NC: Duke University Press, 2006.

Responding to the green economy: how REDD+ and the One Map Initiative are transforming forest governance in Indonesia

Rini Astuti[a,b] and Andrew McGregor[a,b]

[a]School of Geography, Environment and Earth Sciences, Victoria University of Wellington, Aotearoa/New Zealand; [b]Department of Environment and Geography, Macquarie University, Sydney, Australia

This paper analyses the technologies of government that proponents of the Reducing Emissions from Deforestation and forest Degradation (REDD+) mechanism are adopting to influence forest governance in Indonesia. It analyses the aspects of forest governance being problematised; the solutions being constructed; and who is influencing the production and content of these solutions. The research focuses on three aspects of the One Map Initiative: the forest moratorium; forest licensing; and new standards in participative mapping. Our findings show that the initiative has created new opportunities and constraints for forest reform. New disciplinary and participatory technologies have emerged that have created political spaces for activists to actively promote social and environmental justice concerns. However, our analysis also shows tensions for forest stakeholders between engaging in the new opportunities of the green economy and the risk of having political issues rendered technical.

Introduction

'Putting forests at the heart of a green economy' was the slogan of a high-level dinner hosted by the Indonesian REDD+ Agency at the 20th United Nations Conference on Climate Change in Peru, in December 2014.[1] At the event Heru Prasetyo, the Head of the REDD+ Agency, vowed to make REDD+ a reality by the end of 2016.[2] He argued that the REDD+ Agency had delivered the required institutional and governance reforms to begin a system of results-based payments for improved forest carbon management,[3] one of which was the One

Map Initiative (OMI). OMI is a nationwide governance programme that aims to consolidate spatial data in order to develop one integrated geographical information system for Indonesia.[4] In the context of forest governance, data discrepancies in the forestry sector have caused confusion over forest spatial information and contributed to land tenure conflicts and overlapping concessions. The 'green economy' of REDD+ requires a much cleaner and more stable forest landscape to attract carbon investors.[5] OMI aims to create this security by fixing the way forest spatial information is produced and used to govern forests in Indonesia. While critics fear that REDD+ may increase social and environmental injustice, many environmental and Indigenous activists in Indonesia see OMI as providing opportunities to contest the status quo and politicise forest governance in new ways.

REDD+ initiatives, like those in Indonesia, are being unrolled across forested parts of the majority world. Hopes are high that REDD+ will provide a means of addressing forest emissions by creating financial incentives for improved forest management. Pilot projects, however, have proved controversial,[6] with a divide emerging between those seeking to develop practical mechanisms to implement the programme and critics concerned with issues associated with the neoliberalisation of nature.[7] What is becoming increasingly apparent is that REDD+ differs considerably across scale and space, reflecting the differing interests and socio-ecological contexts of forest stakeholders.[8] The novelty of the programme provides it with some malleability in how it is implemented in different places. While it is ostensibly based on neoliberal rationales, in that the production of forest carbon as an economically valuable commodity drives the programme, a variety of other non-economic interests are also involved, such as those seeking biodiversity protection, poverty alleviation, social justice and forest governance reform. As such we see REDD+ as a heterogeneous programme in which the neoliberal discourses inherent in the design of the mechanism encounter others emerging from the particularities of place. What is of interest, then, is who is involved in REDD+, what tactics are being adopted, how forest governance is changing, and with what consequences for different forest stakeholders.

As the leading site for REDD+ finance the experiences of Indonesia with initiatives like OMI will influence how REDD+ evolves in other countries. Our aim in this paper is to analyse how OMI is transforming forest governance in the Indonesian context. We explore how new forms of spatial knowledge are being generated to produce 'governable spaces' for forest carbon investment, and the consequences this is having for forest stakeholders. We are particularly interested in how the processes of producing spatial knowledge provide openings and closures for more progressive forms of forest politics. We adopt a governmentality lens,[9] in order to analyse the new institutions, knowledges and practices that are becoming dominant in forest governance and the role and influence of state and non-state actors.[10] In adopting a governmentality approach we outline what aspects of forest governance are being problematised; what technical solutions are being constructed; and who is influencing the production and content of these solutions.

The paper begins with a brief theoretical review of governmentality and forest governance. We then focus on three key OMI technologies aimed at making forest space more governable – the forest moratorium; One Database; and One Standard. Our argument is that discourses of the green economy are transforming the way spatial knowledge is produced, distributed and stored. The technologies are being used to problematise certain forest governance issues in non-confrontational ways, while avoiding and obscuring politically sensitive problems. In doing so, they seek to generate consensus among forest stakeholders in favour of the green economy. However, we also argue that the processes through which new forms of spatial knowledge are produced are contested, providing new opportunities for non-state actors to shape the production of forest knowledge. As a consequence governmental technologies are outcomes of negotiations of state and non-state actors, and reflect the particularities of place. Thus we argue that neoliberal governmental technologies, such as REDD+, should be seen as sites of struggle, redefining the rules, actors and forms of engagement for all involved in complex and contradictory ways.

Data for this paper were collected as part of a broader three-year study looking at the political ecology of REDD+ in Indonesia. The first author spent 14 months in Indonesia for data collection from 2013 to 2014. We draw on 60 semi-structured interviews and ongoing relationships with government officials, activists, academics and private sector representatives. Data from the interviews are enriched with observations and analysis on related documents and policy archives on forest land use policies.

Governmentality and forest governance

In his conceptualisation of modern forms of governance Foucault coined the term 'governmentality' to refer to attempts to govern human conduct according to widely acceptable 'appropriate' norms and values.[11] Instead of using solely coercive mechanisms, such as disciplinary technologies, governmentality relies on calculated means that are aimed at persuading people to do what is considered good for the population at large.[12] Fletcher uses the term 'environmentalities' to refer to strategies that seek to govern human *and* non-human populations and processes, particularly by regulating human–environment interactions.[13] He identifies four main forms – neoliberal, disciplinary, sovereign and truth environmentalities. Neoliberal environmentality refers to governance through market logics and financial incentives or disincentives that influence or steer the way governed subjects choose to conduct their actions. In contrast, disciplinary environmentalities are pursued through internalising norms and values. Disciplinary approaches aim to produce docile subjects who will act according to shared values and ethics that are considered to be in the best interest of society. Sovereign environmentalities are accomplished through the execution of top-down regulations which establish rules for how people should behave – through fences and fines for example. A final environmentality is exercised in terms of a 'particular conception of the nature and the order of the universe', for instance beliefs in a non-material relationship between humans and nature, as is common in some Indigenous societies.[14] People come to govern themselves in accordance with spiritual beliefs and traditional rules and customs. These competing environmentalities are considered fluid and unstable, intersecting, complementing or even negating each other within particular contexts. They are pursued through a variety of techniques and forms of knowledge, sometimes referred to as governmental technologies, to achieve their aims.[15]

Within governmentality studies government becomes the processes associated with the 'right manner of disposing things'.[16] In order to achieve this goal, authorities devise strategies to influence populations to self-govern their conduct in ways considered convenient to all. Two common sets of practices experts adopt are 'problematisation' and 'rendering technical'.[17] According to Dean, problematisation is a process of identifying defects that need to be fixed.[18] Problems are framed in particular ways to enable 'certain sorts of diagnoses, prescriptions, and techniques'[19] to be proposed as solutions. Effective government requires that these solutions are achievable and non-political – thereby seen as being in the interests of all. This requires empowering particular authoritative but apolitical knowledges and approaches – a practice known as rendering technical.[20] Scott argues that states deliberately generate simplified forms of selected data that will facilitate the effective governance of the population rather than factually representing society.[21] By combining selected data with diverse forms of governmentality, the state is able to depict the reality they represent as a universal truth and hence to outline apolitical and appropriate solutions.[22] Li extends Scott's argument by acknowledging the need to look beyond state simplification to analyse the complexities hidden behind state governance initiatives. By examining the messy assemblages informing state approaches we can generate 'something new – new ways of seeing oneself and others, new problems to be addressed, new modes of calculation and evaluation, new knowledge, and new powers'.[23]

The identification of problems and construction of solutions exemplify 'expertise and constitutes [sic] the boundary between those who are positioned as trustees, with the capacity to diagnose deficiencies in others, and those who are subject to expert direction'.[24] Li warns of the tendency of the trustees to render the domain to be governed non-political as an attempt to take out political-economic concerns and repose them using the 'neutral language of science'.[25] Such attempts privilege those with access to scientific and technical knowledge and marginalise those without it. Li explains further that the process of rendering technical will always be accompanied by practices of confrontation. As we will discuss, this has been the case in Indonesia, where the introduction of OMI as a technical means to resolve long-standing political economic conflicts has engendered oppositional actions among some Indigenous and environmental activists who actively contest and disrupt policy-making processes. However, overt and radical moves opposing OMI are rare; instead contestation takes the form of 'subtle slippages and subversions' oriented at centralising principles of socio-ecological and Indigenous justice within the new technical framework.[26]

This paper draws from the above notion of governmentality to examine the emergence of new technologies aimed at governing Indonesia's forests and forest peoples. We analyse processes of problematisation and rendering technical as well as the responses and strategies of particular forest stakeholders in engaging with them. Thompson et al define forest governance as 'a set of social norms and political assumptions that will steer societies and organizations in a manner that shapes collective decision about the use and management of forest resources'.[27] As a practice, governance is conducted not only by state actors and authorities, it also involves other agents and institutions, incorporating often-incompatible interests. Forest governance attempts to address complex political economic and socio-ecological issues, where forest actors negotiate conflicting interests.[28] It is often administered by state experts who claim scientific knowledge to establish the right to be managers or trustees, rather than entrusting such right to the hands of Indigenous or local communities, whose knowledge is commonly discriminated against.[29] Therefore, transforming Indonesia's forest governance in progressive ways necessarily means challenging the way knowledge is being generated, and whose knowledge counts.

With these insights in mind we now turn our attention to how the REDD+ Agency, backed by the President's Unit for Development Monitoring and Oversight (UKP4), sought to problematise forest governance and render technical particular solutions seen to be in the interests of the Indonesian population.

REDD+ and the problematising of Indonesia's forest governance

Li explains that the usefulness of land for one person partly depends on the exclusion of others.[30] In the context of state control over land the exclusion is legitimised through knowledges, laws and practices that are politically supported by government agencies. These form what Li calls a 'regime of exclusion', that is, a set of mechanisms in which state actors (sometimes in league with capitalist entities) rearrange people and their resources in ways that give privileges to the government to decide how and by whom the resources can be used.[31] Peluso and Vandergeest refer to 'political forests' in explaining the measures taken by the colonial government to dispossess land from Indigenous people and claim it as state land.[32] These measures include technologies of territorialisation, zoning, inscription of law and its enforcement, as well as 'criminalizing many previously common practices'.[33]

During the New Order era Suharto and his allies employed the notion of 'development' to claim some 70% of Indonesia's land as state forest.[34] When a plot of land is assigned a function, as state forest for example, a series of actions follow that change the social and political relationship of the humans that interact with or through that land.[35] For millions of forest-dependent communities whose lands were taken by the state there was little room for dissent. The state employed coercive forms of government, criminalising those who did resist, sometimes denouncing them as communists, which led to social exclusion and made them vulnerable to New Order anti-communist violence.[36] While environmental movements preceded the political changes of 1998, ongoing oppression in the forest sector prevented progress on forest land reforms before then.[37] After 1998, community groups and NGOs became much more outwardly active in pursuing forest rights and justice, albeit with limited success. Indonesia maintains one of the highest rates of deforestation in the world, much of that being driven by large-scale forest concessions provided by the state for oil palm, timber and mining interests.[38]

When REDD+ emerged in 2007 as a possibility for Indonesia, it was derived from broader global-scale concerns about climate change. Greenhouse gas emissions were problematised and REDD+ positioned as one relatively cheap and quick solution that, in adopting an offset mechanism, provided a means to avoid the difficult politics of reducing emissions in wealthy countries.[39] REDD+ architects based their rationale on the assumption that deforestation was being driven by insufficient pricing of the carbon services of forests and provided a solution that incentivised forest protection. In other words, the entrenched political economy of global forest destruction was to be resolved by developing a technical mechanism that would pay people and countries to protect forests.

Despite the limitations of this simplistic approach to forest governance, many academics and environmental and Indigenous activists in Indonesia have seen REDD+ as providing political space to problematise the established forest order. Critics counter that REDD+ facilitates the extension of capitalism's interests,[40] in which social and environmental justice concerns are translated into technocratic prescriptions – in order to enable the transformation of 'disadvantaged groups' (such as Indigenous communities) into 'active citizens' and prepare them to be able to participate in market environmentalism activities.[41] Nonetheless, for activists in Indonesia the opportunity of changing a historically unjust forest system into something more equitable in regards to recognising the land and rights of Indigenous people provides the possibility to 'domesticate' neoliberalism and stimulate social equality. These concerns, tensions and contradictions are rewriting forest politics and governance in Indonesia.

In the following sections we focus on three technologies of forest governance that have emerged as a result of OMI and Indonesia's interests in green economies. In each case problems of forest governance are defined and technical solutions supplied. However, rather than erase forest politics by rendering it technical, we explore how non-state actors and particular government agencies have engaged these technologies and pursued strategies to reflect their diverse interests. Hence by examining this moment of forest reform we expose the politics of rendering technical, showing new technologies are not imposed and untroubled from above but are negotiated through existing political economies of forest governance.

The forest moratorium as a technology of governance

In this section we examine the establishment of a forest moratorium, which involves elements of Fletcher's disciplinary and sovereign environmentalities.[42] The moratorium, a key milestone in Norway's $1 billion commitment to REDD+ in Indonesia, has two key objectives:

> (i) cease licensing in 'primary' forest areas, at least temporarily, in order to dampen high rates of forest loss; and (ii) during this cessation, integrate registries, maps, and regulations concerning the extent and status of licenses and forest cover, to allow for rational forest management.[43]

The moratorium policy renders current practices of forest governance visible and problematises them – implicitly positioning current arrangements as irrational. The policy recognises the need to slow forest loss and to improve forest data if green economies based on forest carbon economies are to be realised. The moratorium was announced through the issuance of Presidential Instruction No 10/2011, which prevents a range of government authorities, including the Ministry of Forestry (MoF) as well as district heads, from issuing new forest licenses. Those that breached these guidelines would be breaking the law. As such the Instruction seeks to govern how government officials and the private sector conduct themselves at various scales of authority, including national, provincial and district levels of government.

To legitimise this substantial shift in forest governance, where the autonomy of government agencies in allocating licences was suspended, REDD+ proponents sought to problematise existing arrangements. The UKP4 and REDD+ Agency criticised the way that forest spatial information was manufactured, distributed and stored by the MoF and other technical ministries or local government agencies. They drew on data discrepancies in the forestry sector that have caused confusion over forest spatial information and led to land tenure conflicts and overlap of concessions. Such discrepancies are common, resulting from the interest of particular actors, both at the district and national level, which profit from retaining the ambiguity of forest spatial information.[44] In its latest study on the overview of Indonesia's forest situation, Forest Watch Indonesia asserts that there were seven million hectares of overlapping forest concessions and a total of 2585 tenure conflicts during 1990–2010.[45] The REDD+ Agency used this sort of information to legitimise the moratorium, cognisant of the need for a cleaner and more stable forest landscape to attract REDD+ investors. Carbon investment is likely to be more successful if it is implemented in a conflict-free landscape where there is clarity over who has the right to the carbon.[46]

The forest moratorium renders problematic both the practices in which forest lands are being governed and the governing institution (the MoF), which is positioned as lacking the political will and capability to implement the moratorium. In one case a tactic used by UKP4 and REDD+ Agency officials was to draw on a map depicting two different versions of primary forests on Papua Island to problematise the asymmetrical coordination and information exchange between different state agencies (Figure 1).[47] One of the maps comes from the MoF,

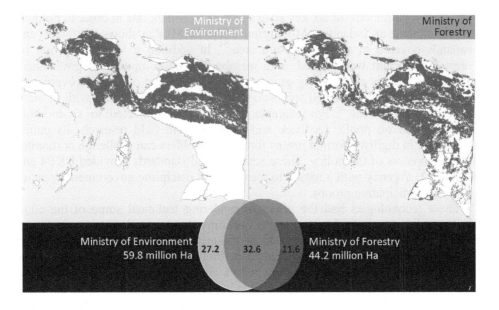

Figure 1   Maps of forest cover in Papua Island.
Source: Samadi, "One Map Movement."

while the other comes from the Ministry of Environment (MoE); these were both presented as official maps in the cabinet meeting chaired by former President Yudhoyono to discuss the implementation of moratorium policy. The maps reflect the degree of uncertainty in Indonesia's forest spatial data and at the same time underlies the more complex problems in its governance. Because of unclear forest boundaries, different methodologies in mapping and inconsistencies in defining what is considered primary and secondary forests, there are at least two different official sets of data pertaining to primary forest in Papua Island. Data from the MoE show 59.8 million ha of forest cover, while MoF data record only around 44.2 million ha.[48]

Having established the problem, the REDD+ Agency and UKP4 positioned the forest moratorium as a technology of government that could overcome the fragmentation of forest spatial information and knowledge.[49] To implement the moratorium required new practices utilising what Scott defines as a process of simplification. That is an attempt to generate a particular reality of the object of governance by showing only information that will support the authority of the governor instead of more complete and complex representations of the issue. The moratorium policy introduces new terminology, 'primary natural forest', in its attempt to technically classify Indonesia's diverse forests into two types: primary natural forest and secondary forest. The first article of the Presidential Instruction states that the implementation of the moratorium will cover only the primary natural forest and peatland. This has caused considerable confusion and contestation over its meaning.[50] For instance, the new terminology has been interpreted by some as meaning primarily undisturbed forests, which means the moratorium covers only a proportion of forests that are already protected under the administration of national park and wildlife sanctuaries.

The moratorium is being implemented and mainstreamed through a range of participative technologies introduced by the REDD+ Agency and UKP4. This includes a requirement for six key state agencies to collaborate in order to produce and disseminate an Indicative Moratorium Map (IMM) that shows the forests and peatlands covered in the moratorium area.[51] In addition a Working Group on Forest Moratorium was formed, with members being composed of government agencies as well as civil society groups. These groups and particular government agencies were to report on 30 technical moratorium activities, each with its standards of achievement.[52] For example, the IMM is subjected to six-monthly reviews based on public feedback and findings from field research. Its public availability in digitised format means forest stakeholders can challenge or monitor the effectiveness of the policy. These activities and standards provided UKP4 and the REDD+ Agency with a means to measure and discipline government agencies and other participating groups.

These technologies had the effect of rendering technical some of the complexities of Indonesia's forest political economy. It obscured sensitive issues of corruption and rent-seeking practices in Indonesia's forest governance. Instead the problems of forest land governance in Indonesia have been constructed as being technical issues, deriving from a lack of comprehensive spatial data and weak inter-ministerial coordination. Following Li we recognise how this practice of rendering technical is generating new methods of governance. By appearing apolitical, outright accusations are avoided and government agencies are able to sit at one table to discuss technical solutions. While NGOs risk de-politicisation by participating in policy making, the formation of these new technical policies can also be 'politically productive as it has the potential to open up new sites of political contestations'.[53] NGOs are using OMI as an opportunity to raise questions about social and environmental justice concerns. Relying on data collected through field verification, for example, the NGO WALHI Central Kalimantan has highlighted the illegal activities of several palm oil plantations that are still clearing forests, despite being included in the areas protected by the moratorium.[54] Such practices use the rules of the moratorium to pursue social justice outcomes.

Despite improved transparency and opportunities for participation in forest governance, civil society has actively contested aspects of the moratorium. NGOs were concerned that it only covers forest under the classification of 'primary natural forests' and peatland, while leaving secondary and disturbed forest unprotected.[55] Activists accuse business-sector lobby groups over the new terminology, which can serve to protect their economic interests. An activist from HUMA, an environmental NGO based in Jakarta, further problematises the moratorium, saying, 'because the secondary forest is not part of the moratorium area, the policy will overlook the social problem caused by tenure conflict that mostly occurred in already allocated forests [for business activities]'.[56] Indeed the moratorium policy is often simplified as a contestation between ecological and economic rationales, while social justice considerations are left aside. This reflects Li's concern that, in presenting technical solutions, policy makers often ignore the social, political and economic factors in designing policies and programmes.[57]

The new practices and policies described above do not guarantee improvements in forest governance. Instead they reveal some emerging strategies in governing the production of forest knowledge in ways that fit the requirements for forest carbon investment. Through OMI, REDD+ proponents have prescribed technocratic approaches to fix spatial truths about forest land use in the hope of generating governable spaces for REDD+ implementation. The production of these spaces entails the adoption of new rationalities of forest governance. For Indonesia to attract REDD+ investment, UKP4 and the REDD+ Agency have attempted to render technical the process of rearranging entrenched power relations within the forest sector. This has involved strategies that ultimately seek to modify and discipline the conduct of state agencies and concessions holders in apolitical ways. It has also provided openings for NGOs to pursue their own politics through participative openings in the technical processes. However, success has been varied. In the following two sections we take a closer look at two other technologies in the context of OMI: the One Database and One Standard technologies.

One Database as a technology of governance

One Database targets the conduct of the private sector, forestry officials and district authorities to make them more accountable for effective green economy investment.[58] One Database enables the state to represent and classify the messiness of Indonesia's forest licences data into an organisation of centralised, audited and standardised knowledges.[59] Li argues that these practices of homogenisation, standardisation and centralisation 'do simplify, but they also generate something new'.[60] In this case UKP4 and the REDD+ Agency officials collect data and facilitate the implementation of One Database pilot projects in three districts. They problematise deficiencies in the governance of forest concessions and propose technical solutions: (1) collection and storage of licences data in one centralised database; and (2) development of a 'situation room', a system of national surveillance that will enable monitoring through real-time satellite data. However, Scott reminds us that experts tend to produce data that only facilitate the connection between problems and their proposed improvements, while obscuring other relevant information.[61] Thus, the improvement projects often risk being compromised to fill the gap between the actual problem and its solution, as we will show to be the case in One Database technology.

Like many REDD+ countries, Indonesia has low governance indices and high levels of corruption in the forest sector.[62] The Anti-corruption Commission, for example, has highlighted 17 loopholes that allow corruption and maladministration of forest land use and land allocation to go undetected.[63] Some of these loopholes are located within the regulations for and mechanisms of issuing forest concessions.[64] The REDD+ Agency has problematised forestry and district officials for utilising loopholes in forest governance – caused by the unavailability of consolidated maps and lack of clear forest boundaries – to facilitate corrupt acts.[65] Consequently multiple concessions are issued for the same area creating hotspots of land contestations and conflicts. As argued at the beginning of the paper, green economy investment, such as REDD+, is likely to be more successful if it is implemented in a conflict-free landscape. Therefore, addressing the messiness of forest licensing governance became a priority for UKP4 and the REDD+ Agency, and was pursued through the One Database technology.

Despite the availability of a set of regulations that governs the mechanisms for issuing concessions, the process of getting one is often seen to rely upon securing favourable terms from close political ties or by bribing government officials.[66] During an interview with the Director of Rimba Makmur Utama (RMU), a proposed REDD+ project in Central Kalimantan, he described the experience of seeking the Minister of Forestry's approval for his project:

> More than four years passed by and I have personally collected 350 signatures required for the concession to be finally reach Minister's table. I did not pay even a cent for under the table transaction [bribe]. It was hard, but, I think it worth every steps that have been taken. It's a green business…I want to do it right and to show that it is possible to get things done without bribery in Indonesia. Everytime someone ask the progress of the concession, I jokingly say 'I think the minister forget to bring his pen everytime he intends to sign it'.[67]

RMU's case has been used by REDD+ proponents to highlight and problematise corruption practices in the forestry sector. In contrast Rimba Raya Conservation, another REDD+ project situated in Central Kalimantan, was widely rumoured to have got the minister's approval thanks to some powerful figures on its board.[68] The institutionalisation of closed-door decision making and corruption among forestry agencies is problematic for green investment.

UKP4 and the REDD+ Agency have initiated the improvement of licensing governance in three REDD+ pilot provinces: Central Kalimantan, East Kalimantan and Jambi.[69] Three districts in Central Kalimantan have been piloted for the programme – Barito Selatan, Kapuas and Kotawaringin Timur. UKP4 and REDD+ Agency officials describe the effort to transform licensing governance through One Database as a radical measure in opening up forest licences data, making it available for public scrutiny. The aim is to prevent discreet decision making in granting concessions and 'under the table' transactions. This is to be achieved through development of an online centralised licensing registry and the establishment of a 'situation room'.[70] The development of the registry starts with the collection and digitation of concession documents and an improvement of archiving mechanisms.

The centralisation of licences data involves a licensing audit reviewing the legal status of concession holders, legal status of permits, environmental sustainability performance, and land tenure and tax/royalty payment compliance.[71] The consolidated registries will make visible the allocation of forestland for private concessions. Thus it will provide a platform for forest governance transparency that enables more informed contestation and critiques. In addition to legal documents, companies are requested to send maps and coordinate points of the concession location to the audit team. The spatial data is used to generate a consolidated map of forest concessions and evaluate the overlaps on areas that are subjected to multiple claims. A REDD+ Agency official explained that in the near future the consolidated map of forest concessions will be overlaid with a map of tenure claims from Indigenous and local communities to assess hotspots of contestations and seek resolutions.[72]

In each district the pilot activity addresses two types of licences: mining and plantation concessions that have been issued both in the state forest area and in forest classified as other land uses. A UKP4 official explained that they chose these two concession types because forest clearance from plantation and mining expansion are the main drivers of deforestation.[73] Most land tenure conflicts are found in these two concession areas. At provincial and national levels the situation room functions as a control room to store a registry of digitised data of concessions and monitor forest changes, such as deforestation or forest fires. The situation room provides a form of centralised surveillance that enables the monitoring of concessionaires' activities in relation to their concessions areas from afar. Using satellite technology providing real-time images, and supported by the provision of a consolidated concessions map, forest monitoring is anticipated to become much more effective. It is expected to reduce illegal logging by prohibiting concessionaires from illegally appropriating forest lands by clearing areas beyond the defined boundaries allocated. The digitation of licensing data provides a system for the local and central governments to track companies' compliance in accomplishing tax and royalty payments. As a set of practices the One Database technology encourages forest stakeholders to self-govern their conduct in ways that fit broader forest sector interests, including those of the green economy.

However, the simple action of collecting concessions documents has proved difficult. UKP4 estimates that only 5% to 40% of the data are readily available at the district level.[74] Most of the documents were not available either because of poor documentation or the intentional action of hiding information. One district official interviewed claimed he was worried that the findings from the audit would put him and his acquaintances in jail.[75] According to UKP4 one of the challenges of collecting concessions documents has been the objection from several District heads to sharing what they classify as 'sensitive' data.[76] Sensitive data usually include illegal concessions issued because of bribery or data documentation that show how officers intentionally overlook noncompliance activities. Discontinuity of data documentation from one administration regime to another is another major obstacle. To ensure the cooperation of the district officials, UKP4 officials attempt to convince district authorities that the findings from the audit process are not going to be utilised for legal action that could criminalise them as a form of compromise.[77] Instead, the rationality of increasing local revenues from plantation and mining tax and royalties is employed to convince the district officials to cooperate in the licences audit process.

The new practices rely on the technical processes of consolidating registries and other spatial information. The registry database produces new forms of knowledge and creates particular truths about forests that will be beneficial for green economy investment. The truths that emerge, however, are compromises negotiated through relationships between state and district officials, tempered by concerns about accessing sensitive information and fears of illegality and prosecution. In promising not to use the data to charge corrupt officials UKP4 and the REDD+ Agency have sought to render technical and depoliticise histories of corruption and injustice. This is seen as the trade-off for the production of more governable forest futures.

One Standard as a technology of governance

A further technology introduced through REDD+ has been the 'One Standard' principle, aimed at governing participative mapping practices, its proponents and their integration in OMI. It does this by standardising how participative maps are made, circulated and used. This section focuses on the background politics that contests such apparently apolitical standards. The government's approach is informed by the necessities of the broader neoliberal environmentalities underpinning REDD+, while opposition is structured around and based on respect for the knowledges of forest communities, making claims based on Fletcher's truth environmentalities. As argued earlier, in order to deliver effective governance, the state will try to render technical its object of governance; however, Scott argues that this practice of simplification is routinely resisted by actors such as NGOs.[78] Li further extends Scott's explanation by highlighting the possibilities of critical engagement, as 'resistance involves not simply rejection but the creation of something new, as people articulate their critiques, find allies, and reposition themselves in relation to the various powers they must confront'.[79] This is the case with the network of Indigenous and environmental activists' resistance toward the government's version of the participative mapping standard. Rather than outright refusal, the network proposes a new standard aimed at promoting the integration of Indigenous spatial knowledge in the state geospatial system as discussed below.

Geospatial law No 4/2011 requires all maps recognised by the state to follow a particular standard to ensure technical compatibility and validity. By integrating maps produced through participative mechanisms, such as Indigenous maps, the process seeks to clarify contestations over land tenure. This is important for REDD+ in Indonesia, as the successful implementation of the new carbon economy relies on its capacity to emphasise its non-carbon benefits, including Indigenous tenure rights. The development of a standard is an attempt by the state to produce a governmental technology to regulate and discipline the conduct of participative mapping proponents.[80] The standard envisions a unilateral cartographic language that will ensure the conformity of participative mapping practices with modern cartographic conventions. It requires the proponents to submit to the practices, and thus values, of modern cartography, the very technology that has previously been spatially capitalised by the state to dispossess Indigenous and local communities of their land and resources.[81] If mapping means asserting claims and controls over particular territory, a standard on mapping is a measure to govern the way claims and controls are to be made. A standard can have both empowering and disempowering effects for marginalised communities. On one hand, it becomes the means of translating the community's land tenure system into spatial information, which will be technically accepted and understood by the government.[82] On the other hand, a technical standard strengthens the modern science domination of Indonesia's state cartography system and risks marginalising traditional or Indigenous spatial knowledges.

The Geospatial Information Agency (GIA), the government agency whose mandate is, among others, to oversee the governance of spatial information in Indonesia, prepared the first draft of the standard during late 2012. The draft consists of four technical documents. The first document explains the procedure of geospatial data provision for participative mapping proponents. The second document regulates the mechanism for participative mapping proponents in updating and adding names of places in the GIA's geospatial system. The third document governs the quality control process for maps produced through participative mapping mechanisms. The last document standardises a mechanism for acquisition and integration of these maps in the One Map system. In addition to the proposed standard, GIA and the REDD+ Agency launched an online application system for participative mapping in late August 2014. The online application theoretically enables every citizen or community group in Indonesia to add spatial information to the basic map. This information can be added directly by the local community through the participative mapping application system and will be followed with data verification by GIA to determine its validity and technical compatibility according to national cartography conventions.

However, participative mapping proponents, such as Jaringan Kerja Pemetaan Partisipatif (Participative Mapping Network – JKPP), have problematised the paradigm of participative mapping developed in the GIA's online application. They have challenged some of the core assumptions within the GIA's approach and have been developing an alternative standard for forest authorities to adopt. In the GIA's standard the meaning of 'participation' is reduced to the involvement of community or other non-state actors in naming places, rivers and lakes, or drawing village boundaries through the online system. As such, one can argue that the GIA's participatory technology has set limited terms for stakeholder engagement. Thus the technology renders the political complexities of Indigenous communities' land tenure claims into bland practices of inputting and digitising spatial information. This perspective differs from some NGOs understandings of 'participative' as a mechanism that carefully involves community members in a set of political processes from the initial planning phase, to the making of the map, to the decision taking as well as to control over the use of the map.[83]

In contrast JKPP proposes a standard in which 'counter-mapping' characteristics are employed as the fundamental paradigm.[84] The standard emphasises the nature of participative mapping as a social movement, through which dispossessed communities can employ mapping as a technology of resistance.[85] JKPP's standard advocates the integration of Indigenous/traditional knowledge in the state's geospatial system by insisting on the existence of traditional knowledge, which has predated the system developed by the state. This of course has political and practical implications – potentially making Indonesia's forest governance more open to social and environmental justice concerns.

In October 2014 JKPP finalised its draft of a participative mapping standard. It is currently under discussion with the REDD+ Agency and GIA. The standard starts by describing the rationales that underpin participative mapping. The rationales emphasise the significant role that Indigenous and local communities have in terms of mapping and spatial knowledge, and the need for this to be acknowledged and represented in the nation's spatial governance system. It also emphasises that the community's possession of spatial knowledge is dynamic and exemplifies the community's tenure rights to their land and resources.

The standard employs several basic concepts originating from counter-mapping principles. For instance, it highlights the role of Indigenous/traditional spatial knowledge as an essential tool for the protection of and struggle for Indigenous tenure rights. The standard addresses the fact that modernism and development – including modern scientific cartography – is a dominant knowledge system that tends to ignore and undermine Indigenous/traditional spatial knowledge.[86] The standard, therefore, proposes that the state's geospatial system facilitate communities to produce their own spatial knowledge; they are referred to in the standard as the 'knowing subject'.[87] By doing this, truth environmentalities are claimed, whereby Indigenous knowledges are valorised and positioned as having a role in governance based on intimate connections with place. Incorporating such approaches only work when there is a mutual acknowledgement of the worth of modern scientific cartography systems and that of traditional spatial knowledge; a basic premise advocated by the JKPP standard.

The standard relies on two philosophical definitions of participative mapping. First, participative mapping is defined as a process of building a collective understanding aimed at the improvement and sustenance of a community's living space. Second, it is also defined as a process of building mutual agreement within the community on the role of participative mapping as a tool to reinforce and strengthen the community's living space. These two broad definitions of participative mapping emphasise how participative mapping can function as a technology of resistance that allow less powerful groups to employ maps as legitimate proof for claiming land and resources.[88] The definitions emphasise social principles as well as technical ones. The social principles require the participative mapping to be initiated from the community's collective agreement, in which the agreement has to be arranged through Free Prior Informed Consent principles. Social data must also be produced to represent the culture, language, Indigenous institutions, tenure system and other social norms.

Despite these innovations, REDD+ investment requires certainty in the form of fixed borders on a map. While this is a necessary step to delineate the community's land from the state or private sector territory, boundaries in an Indigenous spatial system are not exclusively meant as fixed lines that demarcate between two separate compounds.[89] In some Indigenous spatial systems the boundaries are neither static nor fixed but fluid and negotiated.[90] It is a space 'produced through practices related to dwelling, to procuring a livelihood, and through interaction with the environment and they are continually shaped through social relations at multiple scales'.[91] The requirement for the production of fixed boundaries as the condition for legal acknowledgement of a particular Indigenous community, stipulated not only in this standard but also in most other state regulations, has triggered the fabrication of imaginary boundaries and created tensions among communities.[92]

Nevertheless, JKPP's standard challenges the GIA's approach to Indigenous mapping as being a technical exercise and provides an alternative process in which political issues become recognised and embedded in a technical process. It respects and reflects the capabilities and categories of local and Indigenous communities' spatial knowledge and brings counter-mapping and the idioms of tenure rights, power relations and knowledge systems into the governance of Indonesia's land. By engaging with openings provided by REDD+ and struggling for recognition JKPP has attempted to subvert neoliberal rationalities and modern mapping practices to counter and contest the status quo of forest governance. Whether or not JKPP's standard will be effective for countering dispossession of land and for defending Indigenous tenure rights will depend to a great extent upon how forest stakeholders continue to engage with REDD+ opportunities as a means to mainstream the Indigenous/traditional spatial knowledge system. This necessitates a spatial governance system that relies on a capacity to integrate various forms of socio-ecological knowledge systems as well as cultural diversity.[93]

Conclusion

The biggest challenge for the green economy in Indonesia is to transform the messiness of forest governance and produce a 'governable space' for international carbon investment. Through the execution of OMI, REDD+ proponents have sought to bring clarity to confusion over forest land use and allocation. Drawing on three technologies of government – the forest moratorium policy, One Database and One Standard – this paper has argued that the REDD+ Agency and UKP4 have developed strategies aimed at governing the production, distribution and storage of spatial knowledge. In the process they have sought technical solutions to entrenched political problems around corruption, power, land rights, poverty and profit. REDD+ has provided a rationale for restructuring forest governance through the institutionalisation of new forms of knowledge and practice while maintaining an appearance of being non-political and in the interests of all.

Forest stakeholders have responded differently to the technologies applied. In the case of the moratorium many NGOs saw it as an opportunity to advance their interests and participated in the various working groups that were set up. Through participation they were able to contest the limits of the moratorium and expose breaches where companies were clearing protected forest, opening the possibility of prosecution. Local authorities responded much more reluctantly to the One Database initiative, negotiating an amnesty with central government for past breaches of forest laws. By centralising licensing processes and setting up a 'situation room' REDD+ proponents are seeking to reshape their conduct. Finally the One Standard initiative of OMI has been contested by some NGOs who have challenged the basic tenets of the standard and advanced an alternative that promotes Indigenous knowledges. One Standard has provided a channel for more radical interpretations of participative and Indigenous mapping to find an audience among the more conventional cartographers driving OMI.

Our research has shown that the green economy is being pursued through more than just neoliberal environmentalities. Instead elements of disciplinary, sovereign and truth environmentalities are also present in the governmental strategies being adopted by state and non-state actors in Indonesia. We have identified a fertile politics underpinning the production of apparently non-political technical processes. Forest stakeholders are not responding to green economy initiatives in straightforward ways; instead they are exercising agency, strategically engaging in different initiatives to advance their interests. As such initiatives like REDD+ should be seen as sites of contestation, where global priorities encounter diverse political ecologies that shape how programmes unfold. This is shaking up forest governance, rearranging the roles and subjectivities of different actors. Some NGOs have found themselves working alongside state agencies, having access to what was initially inaccessible spatial data and using this to problematise the quality of law enforcement and programme implementation in the forestry sector. District officials have found themselves under increased scrutiny and surveillance from the state while previously oppressed indigenous groups are finding more receptive audiences for their customary land rights claims.

As such the green economy presents new opportunities and risks. NGOs have sought to realise, with some success, the overlap between the interests of carbon investors in governable space with their interests in expanding social and environmental protections. Their new alliances with REDD+ proponents place them in more influential positions than previously. However, risks emerge through the tacit approval these alliances provide for processes that seek to render forest governance a technical issue. By problematising the quality of knowledge, rather than the holders of that knowledge, the green economy avoids histories of corruption and outright conflict with the stakeholders involved. Past injustices are left unaddressed and existing injustices may be prolonged. There is also a risk that, in working with the rationales of the green economy, where social justice issues are often considered 'co-benefits' rather than core priorities, these rationales will reinforce a paradigm and set of technologies that may ultimately work against these interests. After all, there are other ways of producing governable space based on exclusions rather than inclusions – and, as related research has shown,[94] the rules being designed at the national scale don't necessarily translate well across scales.

What we are seeing in Indonesia at the moment, however, are improvements in the visibility of social and environmental justice discourses, as REDD+ proponents in the country engage with civil society and communities, alongside carbon market rationales during the readiness processes. This trend cannot be guaranteed into the future and must be monitored as full scale REDD+ implementation based on reducing carbon emissions moves forward in 2016. Despite progress in mainstreaming values of transparency, participation and Indigenous recognition, there are still many challenges ahead to comprehensively shift forest governance from state processes of land appropriation into mechanisms of land redistribution and re-territorialisation. On paper Indonesia's OMI appears sound: new disciplinary and participatory technologies are emerging to shift the flaws in current forest governance. However, when examined on the ground, many of the technical approaches prescribed by REDD+ proponents rarely suit the political-economic reality of forest governance. Without taking into consideration the diverse political interests shaping forest governance regimes and addressing the conditions these regimes obscure, OMI may yet do little to improve forest governance.

Acknowledgements

The authors would like to thank our research participants and two anonymous reviewers for their insightful suggestions.

Funding

This research is supported by the Marsden Fund Council, from Government funding, administered by the Royal Society of New Zealand.

Notes

1. Under the terms of Presidential Regulation 62/2013, the REDD+ Agency was established in December 2013 to replace the role of the REDD+ Taskforce.
2. Zwick, "Indonesia Vows."
3. Ibid.
4. Samadhi, "One Map Movement."
5. Brockhaus et al., "An Overview of Forest and Land Allocation Policies."
6. Beymer-Farris and Bassett, "The REDD Menace"; and Eilenberg, "Shades of Green and REDD."
7. McGregor et al., "Practical Critique"; and McGregor, "Green and REDD?"
8. McGregor et al., "From Global Policy to Local Politics."
9. Foucault, "Governmentality."
10. Ilcan and Phillips, "Developmentalities and Calculative Practices."
11. Foucault, "Governmentality."
12. Dean, *Governmentality*.
13. Fletcher, "Neoliberal Environmentality."
14. Ibid., 178.
15. Dean, *Governmentality*.
16. Foucault, "Governmentality," 95.
17. Li, *The Will to Improve*.
18. Dean, *Governmentality*.
19. Li, *The Will to Improve*, 7.
20. Li, *The Will to Improve*.
21. Scott, *Seeing like a State*.
22. Ibid.
23. Li, "Beyond 'the State' and Failed Schemes," 389.
24. Li, *The Will to Improve*, 7.
25. Ibid.
26. Stratford, "Micro-strategies of Resistance," 2.
27. Thompson et al., "Seeing REDD+ as a Project of Environmental Governance," 100.
28. Aicher, "Discourse Practices in Environmental Governance."
29. Ibid.
30. Li, "What is Land?"
31. Ibid., 589.
32. Peluso and Vandergeest, "Genealogies of the Political Forest."
33. Ibid., 763.
34. Fay et al., *Getting the Boundaries Right*. 'New Order' was a well-known terminology employed by President Suharto to differentiate his regime from that of the previous president Sukarno's regime.
35. Li, *The Will to Improve*, 5.
36. Bakker, "Who owns the Land?"
37. Peluso, "Whose Woods are These?"
38. Indarto et al., *The Context of REDD+ in Indonesia*.
39. McGregor et al., "Beyond Carbon, More than Forest?"
40. Arsel and Büscher, "Nature™ Inc."
41. Milne and Adams, "Market Masquerades."
42. Fletcher, "Neoliberal Environmentality."
43. Sloan, "Indonesia's Moratorium on New Forest Licenses," 37.
44. Fay et al., *Getting the Boundaries Right*.
45. Purba et al., *Potret Keadaan Hutan Indonesia*.
46. Brockhaus et al., "An Overview of Forest and Land Allocation."
47. Samadhi, "One Map Movement."
48. Ibid.
49. Government of Indonesia, "Letter of Intent."
50. Murdiyarso et al., *Indonesia's Forest Moratorium*.
51. UKP4 and Satgas REDD+, "Laporan Pemantauan Instruksi Presiden."
52. Ibid.
53. Barry, *Political Machines*, 208.
54. Fandi et al., "Melihat Implementasi Inpres Moratorium."
55. Murdiyarso et al., *Indonesia's Forest Moratorium*.
56. Interview, activist 1, 2013.
57. Li, *The Will to Improve*.
58. Samadhi, "Satu Informasi Perizinan."
59. Scott, *Seeing like a State*.
60. Li, "Beyond 'the State' and Failed Schemes," 389.
61. Scott, *Seeing Like a State*.

62. Ebeling and Yasué, "Generating Carbon Finance."
63. KPK, "Integrated White Paper."
64. Dermawan et al., "Preventing the Risks of Corruption."
65. Interview, REDD+ Agency official 1, 2013.
66. Dermawan et al., "Preventing the Risks of Corruption."
67. Interview, RMU Director, 2013.
68. Interview, activist 3, 2013.
69. Samadhi, "Satu Informasi Perizinan."
70. Ibid.
71. Ibid.
72. Ibid.
73. Interview, UKP4 official, 2013.
74. UKP4 and Satgas REDD+, "Laporan Pemantauan Instruksi Presiden."
75. Interview, district official, 2013.
76. Interview, UKP4 official, 2013.
77. Ibid.
78. Scott, *Seeing Like a State.*
79. Li, "Beyond 'the State' and Failed Schemes," 391.
80. Dean, *Governmentality.*
81. Peluso, "Whose Woods are These?"
82. Ibid.
83. JKPP, "Standard Operating Procedures."
84. Ibid.
85. Peluso, "Whose Woods are These?"
86. Harley, "Deconstructing the Map."
87. JKPP, "Standard Operating Procedures."
88. Peluso, "Whose Woods are These?"
89. Roth, "The Challenges of mapping Complex Indigenous Spatiality."
90. Pramono, "Ngekar Utatn Raat Kite."
91. Roth, "The Challenges of mapping Complex Indigenous Spatiality," 211.
92. Astuti and McGregor, "Assembling Indigenous Land Claims."
93. Aicher, "Discourse Practices in Environmental Governance."
94. McGregor et al., "Beyond Carbon, More than Forest?"

Bibliography

Aicher, Christoph. "Discourse Practices in Environmental Governance: Social and Ecological Safeguards of REDD." *Biodiversity and Conservation* 23, no. 14 (2014): 3543–3560. doi:10.1007/s10531-014-0812-5.

Arsel, Murat, and Bram Büscher. "Nature™ Inc: Changes and Continuities in Neoliberal Conservation and Market-based Environmental Policy." *Development and Change* 43, no. 1 (2012): 53–78. doi:10.1111/j.1467-7660.2012.01752.x.

Astuti, Rini, and Andrew McGregor. "Assembling Indigenous Land Claims within the New Political Conjuncture of Forest Governance in Indonesia." *Journal of Peasant Studies*, forthcoming.

Bakker, Laurens. "Who owns the Land? Looking for Law and Power in Reformasi East Kalimantan." Doctoral thesis, Radboud Universiteit Nijmegen, 2009.

Barry, Andrew. *Political Machines: Governing a Technological Society.* London: A&C Black, 2001.

Beymer-Farris, Betsy A., and Thomas J. Bassett. "The REDD Menace: Resurgent Protectionism in Tanzania's Mangrove Forests." *Global Environmental Change*, 22, no. 2 (2012): 332–341. doi:10.1016/j.gloenvcha.2011.11.006.

Brockhaus, Maria, Krystof Obidzinski, Ahmad Dermawan, Yves Laumonier, and Cecilia Luttrell. "An Overview of Forest and Land Allocation Policies in Indonesia: Is the Current Framework Sufficient to meet the Needs of REDD+?" *Forest Policy and Economics* 18 (2012): 30–37. doi:10.1016/j.forpol.2011.09.004.

Dean, Mitchell. *Governmentality: Power and Rule in Modern Society.* Thousand Oaks, CA: Sage, 2009.

Dermawan, A., E. Petkova, A.C. Sinaga, M. Muhajir, and Y. Indriatmoko. *Preventing the Risks of Corruption in REDD+ in Indonesia.* Center for International Forestry Research, 2011. http://www.cifor.org/library/3476/preventing-the-risks-of-corruption-in-redd-in-indonesia/.

Ebeling, Johannes, and Maï Yasué. "Generating Carbon Finance through Avoided Deforestation and its Potential to Create Climatic, Conservation and Human Development Benefits." *Philosophical Transactions of the Royal Society of London B: Biological Sciences* 363, no. 1498 (2008): 1917–1924. doi:10.1098/rstb.2007.0029.

Eilenberg, Michael. "Shades of Green and REDD: Local and Global Contestations over the Value of Forest versus Plantation Development on the Indonesian Forest Frontier." *Asia Pacific Viewpoint* 56, no. 1 (2015): 48–61. doi:10.1111/apv.12084.

Fandi, Aryo Nugroho, and Mariati A. Niun. *Melihat Implementasi Inpres Moratorium Di Kalimantan Tengah: Antara Harapan Dan Kenyataan*. Lingkar Belajar Keadilan Iklim WALHI Kalimantan Tengah, 2013.

Fay, Chip, Martua Sirait, and Ahmad Kusworo. *Getting the Boundaries Right: Indonesia's Urgent Need to Redefine its Forest Estate*. Southeast Asia Policy Research Working Paper 25. Nairobi: World Agroforestry Centre, 2000.

Fletcher, Robert. "Neoliberal Environmentality: Towards a Poststructuralist Political Ecology of the Conservation Debate." *Conservation and Society* 8, no. 3 (2010): 171–181. doi:10.4103/0972-4923.73806.

Foucault, Michel. "Governmentality." In *The Foucault Effect: Studies in Governmentality*, edited by G Burchell, C Gordon and P Miller, 87–104. London: Harvester Wheatsheaf, 1991.

Government of Indonesia. "Letter of Intent between Government of Republic of Indonesia and Government of Kingdom of Norway on the Cooperation on Reducing Emissions from Deforestation and Forest Degradation." Jakarta, 2010.

Harley, J. B. "Deconstructing the Map." *Cartographica: The International Journal for Geographic Information and Geovisualization* 26, no. 2 (1989): 1–20. doi: 10.3138/E635-7827-1757-9T53.

Ilcan, Suzan, and Lynne Phillips. "Developmentalities and Calculative Practices: The Millennium Development Goals." *Antipode* 42, no. 4 (2010): 844–874. doi:10.1111/j.1467-8330.2010.00778.x.

Indarto, G. B., P. Muharjanti, J. Khatarina, I. Pulungan, F. Ivalerina, J. Rahman, M. N. Prana, Ida Aju Pradnja Resosudarmo, and Efrian Muharrom. *The Context of REDD+ in Indonesia: Drivers, Agents, and Institutions*. Bogor: Cifor, 2012.

JKPP. "Standard Operating Procedures: Penyelenggaraan Pemetaan Partisipatif Dan Pengendalian Kualitas Peta Partisipatif." Jaringan Kerja Pemetaan Partisipatif, 2014.

KPK. "Integrated White Paper: Semiloka Menuju Kawasan Hutan Yang Berkepastian Hukum Dan Berkeadilan." Komisi Pemberantasn Korupsi, 2012.

Li, Tania Murray. "Beyond 'the State' and Failed Schemes." *American Anthropologist* 107, no. 3 (2005): 383–394.

Li, Tania Murray. *The Will to Improve: Governmentality, Development, and the Practice of Politics*. Durham, NC: Duke University Press, 2007.

Li, Tania Murray. "What is Land? Assembling a Resource for Global Investment." *Transactions of the Institute of British Geographers* 39, no. 4 (2014): 589–602. doi: 10.1111/tran.12065.

McGregor, Andrew. "Green and REDD? Towards a Political Ecology of Deforestation in Aceh, Indonesia." *Human Geography* 3, no. 2 (2010): 21–34.

McGregor, Andrew, Edward Challies, Peter Howson, Rini Astuti, Rowan Dixon, Bethany Haalboom, Michael Gavin, Luca Tacconi, and Suraya Afiff. "Beyond Carbon, More than Forest? REDD+ Governmentality in Indonesia." *Environment and Planning A* 47, no. 1 (2015): 138–155. doi:10.1068/a140054p.

McGregor, Andrew. "Michael Eilenberg, and Joana Borges Coutinho. "From Global Policy to Local Politics: The Social Dynamics of REDD+ in Asia Pacific". Asia." *Pacific Viewpoint* 56, no. 1 (2015): 1–5. doi:10.1111/apv.12091.

McGregor, Andrew, Sean Weaver, Edward Challies, Peter Howson, Rini Astuti, and Bethany Haalboom. "Practical Critique: Bridging the Gap between Critical and Practice-oriented REDD+ Research Communities." *Asia Pacific Viewpoint* 55, no. 3 (2014): 277–291. doi:10.1111/apv.12064.

Milne, Sarah, and Bill Adams. "Market Masquerades: Uncovering the Politics of Community-level Payments for Environmental Services in Cambodia." *Development and Change* 43, no. 1 (2012): 133–158. doi:10.1111/j.1467-7660.2011.01748.x.

Murdiyarso, Daniel, Sonia Dewi, D. Lawrence, and Frances Seymour. *Indonesia's Forest Moratorium: A Stepping Stone to Better Forest Governance?* Bogor: Cifor, 2011.

Peluso, Nancy Lee. "Whose Woods are These? Counter-mapping Forest Territories in Kalimantan, Indonesia." *Antipode* 27, no. 4 (1995): 383–406. doi: 10.1111/j.1467-8330.1995.tb00286.x.

Peluso, Nancy Lee, and Peter Vandergeest. "Genealogies of the Political Forest and Customary Rights in Indonesia, Malaysia, and Thailand." *Journal of Asian Studies* 60, no. 3 (2001): 761–812. doi: 10.2307/2700109.

Pramono, Albertus H. "Ngekar Utatn Raat Kite: A Look into Cartographic Encounters in Counter-mapping Exercises in West Kalimantan, Indonesia." University of Hawai, 2013.

Purba, Christian P., Soelthon G. Nanggara, Markus Ratriyono, Isnenti Apriani, Linda Rosalina, Nike A. Sari, and Abu H. Meridian. *Potret Keadaan Hutan Indonesia Periode 2009–2013*. Bogor: Forest Watch Indonesia, 2014.

Roth, Robin. "The Challenges of mapping Complex Indigenous Spatiality: From Abstract Space to Dwelling Space." *Cultural Geographies* 16, no. 2 (2009): 207–227. doi:10.1177/1474474008101517.

Samadhi, Nirarta. "One Map Movement." Paper presented at the Semiloka: Menuju Kawasan Hutan yang Berkepastian Hukum dan Berkeadilan, Jakarta, December 2012.

Samadhi, Nirarta. "Satu Informasi Perizinan: Menuju Integrasi Lintas Sektor Dan Daerah Untuk Pelaksanaan Pembangunan Yang Lebih Akuntabel." Paper presented at the Rapat Pembekalan Nasional Tata Kelola Pemerintahan, Jakarta, 2014.

Scott, James C. *Seeing like a State: How Certain Schemes to Improve the Human Condition have Failed*. New Haven, CT: Yale University Press, 1998.

Sloan, Sean. "Indonesia's Moratorium on New Forest Licenses: An Update." *Land Use Policy* 38 (2014): 37–40. doi:10.1016/j.landusepol.2013.10.018.

Stratford, Helen. "Micro-strategies of Resistance." *Resources for Feminist Research* 29, nos. 3–4 (2002): 223–232.

Thompson, Mary C., Manali Baruah, and Edward R. Carr. "Seeing REDD+ as a Project of Environmental Governance." *Environmental Science & Policy* 14, no. 2 (2011): 100–110. doi:10.1016/j.envsci.2010.11.006.

UKP4, and Satgas REDD+. "Laporan Pemantauan Instruksi Presiden No. 10/2011: Hasil Capain Dan Tindak Lanjut." Kelompok Kerja Monitoring Moratorium Satuan Tugas REDD+, 2013.

Zwick, Steve. "Indonesia Vows: We will be Ready for Results-based REDD Payments by End of 2016." *Ecosystem Marketplace*, 2014. http://www.ecosystemmarketplace.com/articles/indonesia-vows-will-ready-results-based-redd-payments-end-2016/.

The neoliberalisation of forestry governance, market environmentalism and re-territorialisation in Uganda

Adrian Nel

Department of Geography, University of Kwazulu-Natal, South Africa

There is often a disjuncture between idealised forestry governance models which posit a 'win-win for community and environment' through participatory, multi-stakeholder international development discourses and interventions – and the actually existing processes and structures of natural resource government through which they are articulated. By applying, first, established theorisations of the initial territorialisation of state forestry territory, then conceptualisations of re- and de-territorialisation, derived from Deleuzo-Guattarian formulations, this paper expands on post-structuralist lines of inquiry on the political ecology of forestry to explore substantive transformations in forestry governance in Uganda. It specifically details the role that market environmentalism – the extension of market mechanisms, including carbon forestry, to natural resource governance – plays in reorienting assemblages of actors engaged in forestry governance and in changing configurations of state forestry territory.

Introduction: assembling a human–natural regime over time

Post-structuralist perspectives in political ecology have increasingly fed into the nuanced theorisation of the neoliberalisation of nature,[1] and in particular assemblage approaches have influenced theorisations of socio-natural relations in the Anthropocene: a historical–geographical conjuncture that signals the end of the presumption of distinct human and natural worlds.[2] Today the scope and scale of human influence pose both new global threats, notably climate change and biodiversity loss, and exigencies for multi-scalar human interventions, particularly concerning forestry. While there has never been a shortage of deforestation narratives in countries of the global South such as Uganda,[3] nor of governmental projects that emphasise the 'urgent imperative' of interventions,[4] these have intensified over the past decade, in step with the emergence of transnational market environmentalism. The ideology underpinning market environmentalism seeks to extend market mechanisms to natural resource governance and encourage the mitigation of both climate change and biodiversity loss through initiatives such as carbon forestry.

The challenge for theorists, in response to the emergence of initiatives as market environmentalism, has been to set out these new practices and the various transformations and rearrangements of laws, governance, capital, discourse and expertise that attend them. Further, documenting the gap between the idealised governance models and the actually existing, historically situated socio-natural assemblages they engage can produce interesting insights about forestry regimes. To this extent political ecology approaches look beyond multi-level 'environmental governance',[5] in order to emphasise the unevenness of change and the historical, spatial, material spaces and places in and through which the spatial relationships between societies, economies, the state and its forestry and conservation territories continue to unfold.[6]

Academics such as Murray Li and Descheneau and Patersen have extended assemblage accounts into explorations of both virtual and material forestry and carbon markets spaces.[7] However, the discussion of territory and processes of territorialisation in relation to actually existing carbon forestry accounts is under-emphasised, particularly regarding the ways carbon sequestration 'takes place' in particular country contexts. There is significant conceptual scope for engagement with territory in an assemblage approach, given the centrality of movements of territorialisation and re-territorialisation in the assertion of an assemblage's spatial identity. A focus on territory is certainly not lacking regarding state forestry in general, however. Smith's conception of the 'production of nature' and Vandergeest and Peluso's account of the territorialisation of state forestry resources are particularly instructive and influential in this regard.[8]

In this article I consider the relationship between forestry governance and forestry territory in Uganda. I first draw from a classical approach that seeks to describe the initial territorialisation of forest resources as it relates to the emergence of the colonial state. This account serves as a backdrop to the further discussion of later transformations, as the *de jure* spatial dispensation established during this phase was to endure to the current moment. Then, more experimentally, I engage with assemblage accounts of territory to explain contemporary changes enwrapped in the localisation of the transnational interventions that constitutes market environmentalism. I characterise these changes, which include the inculcation of carbon forestry and the neoliberalisation of the forestry sector, as two interconnected trajectories of re-territorialisation and de-territorialisation. I use this distinction to highlight qualitative changes in the intentions and outcomes of governance change.

The aim is to extend the theorisation of assemblage and territoriality to a consideration of the way emergent forestry governance assemblages in the Anthropocene relate to state forestry territory. In so doing, I provide an experimental account of actually existing market environmentalism in Uganda, in relation to which individual forestry interventions might be evaluated, and against which more progressive arrangements might be envisioned. As Benterrak et al put it, our concerns should shift from representational thinking – the differentiation of things in hierarchies – to compositional thinking.[9] The questions those concerned with 'management' should ask are not about stratifications, technocracy and divides between human and protected 'natures' (a representational approach). Rather, questions should be posed about the kinds of communities of humans and non-humans, of people and trees we imagine, particularly where forestry contexts are contested, as they are in Uganda.

In addressing this context I draw from interviews with players within government agencies and NGOs conducted during my PhD fieldwork in Uganda in 2012. While I heuristically adopt the often used vertical imagery of the state as 'above' non-state actors, and discuss processes of upscaling and 'downscaling' in this paper,[10] the focus on assemblage and territory affords a topological, or ontologically 'flat' view of networked governance. This is useful in considering how the relations of power play out in interactions of territorialised assemblages.[11] A historically situated and actor-centred account of government as an assemblage of distributed agencies – which 'reach out' across space to engage each other – engages the incentives and constraints felt by state and non-state actors, and organisational capacities, and focuses on the disjuncture between idealised governance models (which posit a 'win-win for community and environment'), and the actually existing processes, territories and structures of natural resource government through which they are articulated.[12] I begin by setting out an account of assemblage and territoriality.

Assemblage, territory and state forestry in Uganda

An assemblage is a collection of heterogeneous elements or entities, as well as processes and practices, that are organised and cohere together, and which are inherently spatial and territorial.[13] As Delaney puts it 'territorialising projects give social–material expression to ideologies [and] territory reflects ideological conceptions of space and power'.[14] There is a reflexive link between territory and social forms such as assemblages for, depending on how a given social order is organised 'certain territorial expressions will be possible and more or less serviceable and others will be less likely'.[15] Thus, in order to understand carbon forestry and market environmentalism in Uganda, we need to understand that the territoriality of the Ugandan 'forest estate', as it is called, and the hierarchical, territorialised governance form that oversees it, are both co-produced. In this light hierarchical state forestry governance formations (and forestry governance in Uganda in particular[16]) can usefully be characterised as assemblages – and moreover as arborescent territorialised assemblages – which promote order and hierarchy and make claims to jurisdiction over social space. In this framing claimed social spaces, and forestry spaces in this instance, are articulated through processes of territorialisation. Territorialisation and re/de-territorialisation are movements between which 'everything happens' as the spatial footprint and identity of an assemblage is stabilised and destabilised respectively, and as heterogeneous parts come together and come apart.[17]

Uganda provides a particularly interesting context for theorisation; it is termed a 'funny place to store carbon' because of the challenges Reducing the Emissions from Deforestation and forest Degradation (REDD) and similar interventions face there.[18] According to Turyahabwe and Banana, despite regular reformulations and revisions it remains unclear whether forestry policies, laws and territories are acceptable to the local people and appropriate to the local situation,[19] let alone the interventions that overlay them.[20] In this regard it is important to grapple with Uganda's 'socioecological question', which pertains to how the state has attempted to mediate – often through violence, and largely unsatisfactorily (in ways that privilege capital accumulation) – tensions between humans and the 'natural environment' in the forestry territories of Uganda.[21]

Neoliberal forestry regimes emerged through initial state forestry management and have evolved to include new non-state actors, including NGOs, corporations in the private sector and multilateral institutions, in what Swyngedouw has termed 'governance-beyond-the-state'.[22] Here, non-state actors take on governance roles in a transnational hierarchy as a way to attend to the 'socioecological question' in places such as Uganda. In doing so, central managerial control of forest territories and 'resources' is unevenly reasserted and can even be undermined. Such emergent structures of governance-beyond-the-state have been characterised as neither national nor global, part public, part private assemblages, with bits and pieces of territorial infrastructure, legal rights and institutional authority.[23] As unstable power formations in the making, these assemblages are implicated in a transformation of politics that restructures the power relations between state and non-state actors. Importantly this happens in ways that do not diminish state sovereignty and planning capacities.[24] Rather, the state comes to work in concert with new actors involved in governance, through calculative practices such as carbon forestry that reshape the rural and legitimise state authority in rendering climate change and biodiversity loss governable.[25] Before considering such contemporary re-assemblings and re-territorialisations, however, I draw back to consider the initial territorialisation of forestry territories in Uganda.

Territorialisation

Territorialisation as a concept broadly relates to behaviours related to the establishment and defence of territories as bounded social spaces with insides/outsides. It is the realigning of interactions between power, bounded space and meaning that makes practices territorial or territorialising. Territorialistion includes: the demarcating or partitioning of space; the classification of spaces as territories; communication that a territory exists and where the boundaries are; communication about the conditions and consequences for entering or leaving a territory; and following through with implicit or explicit threats when conditions are not followed.[26] The map in Figure 1 shows what is today considered the protected forest estate in Uganda (although my conceptualisation of forest territory includes forest 'resources' on so-called private and communal lands).[27] While it would be impossible to scope the extended history of forestry itself in Uganda, in this section I parse a brief history of the primary processes and events that inculcated these *de jure*, mapped forestry spaces.

The initial emergence of forest territories in the early 20th century was linked not only to the expansion of imperial capital, but also to the internal territorialisation and normalisation of the nascent colonial protectorate (1894–962), which was engaged with the dilemma of a stabilising and consolidating foreign minority rule over an indigenous majority. This required a regime of 'differentiation and institutional segregation', or 'indirect rule', which amounted

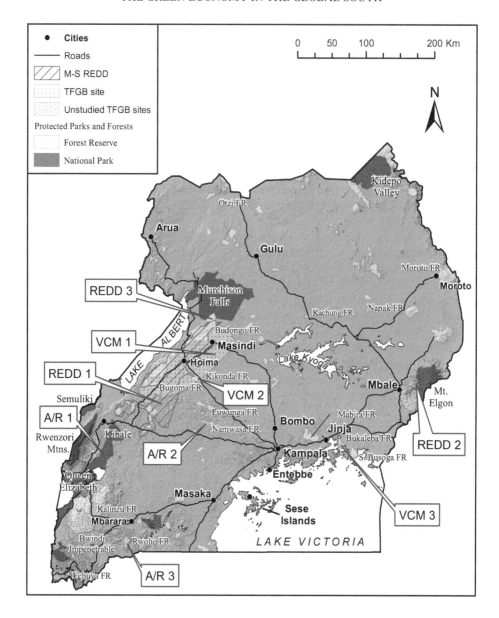

Figure 1. Protected areas, including National Parks and Forest Reserves.
Notes: What I term the forest estate is comprised of the Permanent Forest Estate (PFE), which includes Central Forest Reserves (green) and forested National Parks, as well as forests on 'private and community land' (un-pictured). Carbon forestry project sites for Reducing Emissions from Deforestation and forest Degradation (REDD+), Afforestation/ Reforestation Clean Development Mechanism (A/R CDM) and Voluntary Carbon Market (VCM) projects are also indicated. Notable large projects include the prospective Murchinson Semuliki (M-S) REDD project, and the Trees For Global Benefit (TFGB) VCM project. Source: Brice Gijbertsen, Cartographer, University of Kwazulu-Natal.

to decentralised despotism and entailed the disruption of pre-existing socio-ecological controls over local socio-natures.[28] Cavanagh and Himmelfarb describe the emergence of the Uganda protectorate as an 'ecological project' and the colonial state itself as a predatory socio–ecological system, designed to finance the costly, and contested, process of imperialism through the capture, via taxation, of 'un-useful subsistence peasantry' into export-oriented cash crop production.[29]

Colonialism radically changed the relationship between state and 'society', according to Mamdani, abstracting law making, property systems and ecological control from the societies within which they were embedded, and centralising them within the 'modern' state.[30] This predation was often violent and had massive implications for local populations, disrupting systems of patronage and environmental control,[31] resulting in a calamitous increase in trypansomiasis (sleeping sickness) and bovine death, and 'East Coast Fever'. Doyle, for instance, describes the role of colonial conquest – in the form of a vendetta by one Lord Lugard against the militant Bunyoro–Kitara king Kabalega – in effectively bringing about the conjoined destabilisation of ecological and social systems in the area, which, when compounded by disease, was to initiate a lasting demographic decline.[32]

Vandergeest and Peluso talk of the under-emphasis in critical research on internal territorialisation in state making and the establishment of control over resources and the people using them, and of the importance of 'abstract space', in this case 'forest' territories to the territoriality of the modern state.[33] This abstract space is linear, can be cut into discrete units and homogenised to be represented as uniform and equivalent in any given territory. Scott also considers this form of abstraction in unpacking how states historically transformed 'forests' from nature to 'natural resource' in order to render them governable, reducing complex habitats in order to inscribe principles of commercial extraction with attendant soil depletion and epidemics resulting from the abstract logics applied.[34] In Uganda the British protectorate administration declared most land in the territory 'Crown land' under the Crown Lands Ordinance 1903, where rights of those communities who inhabited these lands were transferred from clan heads to chiefs, and were lost or could be exercised at the pleasure of the colonial governor. Similar declarations occurred elsewhere.[35] Part of this process of abstraction found its advancement in the Forests Ordinance, enacted in 1913 by the Uganda Protectorate. This entailed the enclosure of large occupied areas of the hitherto common lands (as colonial tenure systems were still taking their hold), including grazing lands, community forests and grasslands.[36] In this way, first through 'the stroke of a pen',[37] and then with the use of force, long-standing rights of possession, use, and stewardship were rendered illegitimate.[38]

Formal territorialisation

The process of territorialisation involved the formalisation of enclosure through the gazetting of the Forest Reserves from the 1930s onwards. This was propagated by an emergent, hierarchical forestry management institution, the colonial Uganda Forest Department (UFD), which was established in 1889 and was active until 1961, although it continued under the same name in the immediate post-independence period. I will later describe how a reformed version of this institution came to occupy the centre of what could be characterised in Deleuzo-Guarttarian terms as an *arborescent*, or tree-like assemblage; one which seeks to promote centralised order and hierarchy (in contrast to 'horizontal', spontaneous *rhizomatic* assemblages). Prefiguring this, the UFD itself was characterised by Turyahabwe and Banana as a 'highly regulatory limited stake-holder colonial forest service'.[39]

Formal territorialisation occurred for numerous reasons, including for unsustainable hardwoods extraction and, as Neumann describes it in Uganda's case, through the use of tsetse fly control measures as a justification for enclosure.[40] This extended phase essentially solidified the spatial distribution of the 'forest estate' – which I define as including both designated protected areas and forests outside of these – both materially and expressively. The former related to boundaries, markers and the enforcement of new social relations of production; the latter to hegemonic claims to legitimacy through the codification in law of the status of individual forests reserves.

The process of formal territorialisation was to continue over the years, with impetus for protection stemming from the awareness of the need for water tower protection in areas like Mount Elgon in the East and the Albertine Rift in the West, which feed the Great Lakes, and for agriculture, after the construction of the East African Railway, with the rising importance of coffee production to the state. Individual agency was also to play its part, in the enclosure of national parks around the 1960s, with individuals such as Cowie and Struhsaker advocating the active conservation of wildlife and tropical high forests.[41]

The use of scientific knowledge was also to contribute to managerial conservation and terrestrial land management practices through the process of territorialisation, and was instrumental in proscribing colonial management practices (the influence of Langdale Brown's evaluation of biodiversity types is a prominent example[42]). Accordingly Webster and Osmaston's accounts of the history of the forestry sector (most recently the 2003 edition chronicling 1951 to 1965) chart extensive processes of mapping and enumeration, infrastructure development and expenditure, management, research and administration during the stewardship of the UFD.[43] However, these constructions are partial and, while managerialist perspectives reflect ideal conservation landscapes, these do not necessarily pertain in practice – particularly where the territoriality of protected forest areas is contested. In his preface to Webster and Osmaston's *History of the Uganda Forest Department 1951–1965* (a period of relative stability and success for the UFD, with high forest regeneration and plantation establishment), John Hudson remarked, the 'difficulties and uncertainties described in this book are still faced by foresters today'. Encroachment and theft persist as "problems" for forestry managers, and demarcating the boundaries to aid detection and control is as costly and logistically difficult today as they were 40 years ago.[44]

Contested forestry territory, diverse agency and power

What ongoing difficulties in governance point to is that processes of enclosure and territorialisation are never complete and are continually contested. The protected forest estate is at times more concretely visible in the maps and policy documents of state than it is materially actualised; as on many individual forest territories reserves conceptions of space do not translate into concrete instantiations of 'protected' places. While territories are thus at times anchored in abstraction, and are linked to hierarchy and power relations at the national scale, the hierarchical, and its refusal, is anchored in the everyday through multi-faceted local level resistances that came into friction with the nascent state's territorialisation of forest resources.[45] In this sense, processes of territorialisation are always countered by movements of de-territorialisation in Uganda; there can be alternative narratives, signs and discourses specific to particular places, which are more than mere 'resistance', as they undermine state managerial and project claims to space and territory itself in favour of more local, community-oriented territorialisations.[46]

In looking beyond the state, then, I depart from a sole reliance on state territoriality to begin to approach the Deleuzo-Guattarian sense of a territorialised assemblage, exploring diverse agencies within forestry governance, and the oscillating centralisation and decentralisation of power over forestry territory. It can be asserted that forest governance has in many respects long been a multilateral process, through the influence of donors and international NGOs after independence, and before, through the influence of colonial societies like the Society for the Preservation of the Wild Fauna of the Empire (SPWFE – now Flora and Fauna International).[47] This was linked to movements of centralisation and decentralisation both before and after Independence in 1962.[48] The state's capacity for centralised control during the colonial period for instance waxed and waned as the colonial state's revenues and manpower were taxed by other imperialist activities and wars in Europe, necessitating local control.[49] However, the local government forestry service and its Local Forest Reserves (LFRs), administered under the kingdoms, was centralised by 1967 when the country's republican constitution was adopted. By 1993 the government had adopted a formal policy of decentralisation in line with prevailing sentiment following the Rio Convention on Biological Diversity and 'Local Agenda 21', which envisaged a community-led response to sustainable development challenges locally. However, over 20 years later 'friction between the central government Forestry Department and local governments continues because the laws governing local governments and those governing forestry have not clarified whether forests are decentralised or not'.[50]

The process of territorialisation inscribed sets of power relationships which linked to movements of centralisation and decentralisation, and these were to have a lasting legacy for both forestry and Ugandan politics in the postcolonial period. When the new NRM regime, and its leader Yoweri Museveni (a 'Banyankole' from the Southwest) came to power in 1986 (overthrowing Milton Obote's second regime), it faced a legacy of state disintegration, deep wounds left by political violence and involuntary displacement since the time of Idi Amin, which had exacerbated pressures on Forest Reserves through encroachment after Amin's decree of 'double production'. The young government was eager to make its presence felt to its populace, in part through a massive recentralisation of power over forestry resources. With initial donor-funded support it engaged in the reassertion of the protected area territories, which included evictions of so called 'encroachers' from the said areas, many of whom had been resident for decades.

Whereas Museveni won favour from the West in the mid- to late 1990s for such activities, and for his actions in securing the country under the NRM government, political structures which had been built on ethnicity were retained, particularly the 'ethno–military' complex.[51] This came at the expense of a common national identity and national reconciliation, conceptions of citizenship and a sense of belonging and solidarity, so easily dispensed with during evictions, perpetuating deep rooted divisions.[52] In the absence of effective institutions of forestry governance, and in combination with the confluence of a legacy of state violence, conflict, forced and unforced migration (both within Uganda and across borders), unequal power relationships and pressures on the forest estate manifested not as an unfortunate outlier to otherwise effective management, but as a central, ongoing tension within forestry governance.[53]

Re-territorialisation

To engage with the contemporary moment in Ugandan forestry I would like to characterise how the post-2000 period in Uganda involved a process of re-territorialisation, and the emergence of an assemblage of actors comprising what Swyngedouw terms 'governance-beyond-the-state'.[54] I contend that this assemblage and its leaning towards 'flow based' governance only gives selective expression to the forest estate, privileging public–private partnerships and private sector planting in policy and institutional reforms to accomplish a partial de-statisation of former state domains.[55] In so doing it entails a re-crafting of communal or so-called 'private forestry' to align with market agendas.

To understand how this comes about we need to understand re-territorialisation as a change to forestry territoriality, facilitated through processes of rescaling that mediate resource agendas and hierarchies.[56] These rescalings include what might heuristically be called the upscaling of authority and agenda setting (to 'higher' forms of regulation 'above' the state), downscaling (to project- or ecosystem-level interventions) and outscaling (through privatisation, deregulation and decentralisation). Importantly such rescalings and the re-territorialistions they evoke are not mutually exclusive to what I will describe as de-territorialisation, but are part of the same trajectory of change.

The effect of these changes is to alter how resources, land use and nature are managed, changing the role of the state such that 'nature is no longer a national resource or instrument of national planning and production; it is something to be governed, consumed, and marketed – locally and globally'.[57] With regard to the diverse agencies engaged in governance, while non-state authorities had been involved in forestry governance for decades, the momentum underpinning their involvement and the logics of rescaling have intensified in line with the historical–geographical conjuncture of the Anthropocene. Where the global community is increasingly concerned about a changing climate, carbon forestry projects represent a key feature of global mitigation efforts against climate change.

Upscaling

The trajectory of upscaling of forestry governance occurs through the renegotiation of decision-making authority, or perhaps less strongly the seceding of agenda-setting capacities to supra-state entities like the World Bank's Forest Carbon Partnership Facility (FCPF), and to donors. In Ugandan forestry governance the unfolding of this process has come to be directed around concerns about sustainability, community development and climate security, where aid and flows of climate finance have taken on a direct security role to privilege public–private contractual networks in the assemblage of governance-beyond-the-state that emerges.[58]

Concerning the development of the National REDD readiness proposal for the FCPF, for instance, while the National Forest Authority (NFA) (and by extension the Ministry of Water and Environment – MWE), was the official Designated National Authority, it could only prepare Terms of Reference, sanction payment and receive reports in line with FCPF guidelines. Similarly, when it came to the development of specific modalities in the preparations, the Norwegian development agency, NORAD, the primary donor, did not want to give the NFA money directly following public corruption scandals within the institution in 2009. NGOs were contracted to accomplish the set tasks: the International Union for Conservation of Nature (IUCN) was contracted to do the consultation strategy (with the Bennet and Batwa peoples only – those defined as 'indigenous' in the country) and the charity CARE to prepare a grievance strategy. In such instances, as the former head of the National REDD steering committee put it, 'he who pays the piper calls the tune'.[59]

I use the term 'upscaling' heuristically, for geographers have critiqued hierarchical or horizontal imagery of state power, but to Allen and Cochrane topological thinking presents a nuanced understanding that 'the powers of the state are not so much "above us" as more or less present through mediated and real time connections, some direct, some more distanced...what states possess is reach not height'.[60] The concept of reach implies that the state 'draws' others actors, including donors, civil society and private sector actors (even those engaged in the rural extremities of the country), into negotiation with it. The Uganda Climate Change Unit (CCU) and the National REDD Steering Committee, both housed within the MWE perform these reach functions. Although these nodes may be small, they are powerful and have mediator roles in regard to the claims to territory and legitimacy brought to bear in the lodging, negotiation and accommodation of different actors in the assemblage. Thus, while NGOs and multilateral actors take on governance roles, the state retains some significant degree of coordination over their activities.

Through the intervention of supranational actors such as donors, local institutions are reconfigured in conjunction with the state, which nevertheless remains relevant and does not disappear from neoliberal regimes. Accordingly the most overt drawing together of the 'actually existing neoliberalism' of forestry governance in Uganda was through a donor initiated five-year process – entitled the Forest Resources Management and Conservation Programme (FRMCP) – which gave effect to the country's forestry sector reforms from 1999 to 2003 through the privatisation and deregulation of forestry, and the creation of a quasi-parastatal forestry authority. This was funded through a €12 million 'basket' of funds described in the final section. The reforms sought to disassemble and split up the UFD because of its 'clear weaknesses', including corruption, 'rampant deforestation' and 'many failed initiatives to expand the plantation resource'.[61]

In keeping with the spirit of prior Structural Adjustment Programmes (1987), the reforms entailed the dissolution of the UFD into a streamlined, semi-privatised forestry 'authority' with ancillary bodies to support it, the aim being to take it 'out of politics', although this was set to later fail. There were three new forestry institutions, housed under the MWE, which reconfigured the jurisdiction over the forest estate between them. These institutions included a semi-autonomous, 'efficient' NFA for the management of 15% of the forest estate codified as 'Central Forest Reserves' (CFRs); a District Forest Service (DFS) to be incorporated into local government to manage what at the time was 70% of the forest estate in 'local forest reserves' and reserves on 'private lands'; and a 'streamlined' oversight office (a 'Government of Uganda watchdog'), which entitled the Forest Sector Support Department (FSSD) to oversee coordination in the sector.

The final 15% of the forest estate had previously been incorporated into National Parks under the ambit of the newly created Uganda Wildlife Authority.[62] Additional offshoot initiatives to complete the institutional configuration included the African Development Bank and the Norway-funded FIEFOC project and the SPGS, bodies which Khisa termed governing non-governmental organisations (GONGOs).[63] The 2003 National Forestry and Tree Planting Act (NFTPA) was additionally legislated under the advice of this secretariat.

During the reform process it is said that proponents and opponents of the reforms clashed over various issues, apparently long into the night on at least one occasion. The most contentious issues were the splitting of the commercial and regulatory wings of forestry into separate institutions, and the decentralisation of the District Forest Estate, which required its separation from the NFA. The proponents of the reform won through on the claim to parliament and the donors – the EU and Norway – that the NFA would be self-sustaining and would operate without central treasury funding, which was a key concern for parliamentarians. It was to do this through licensing and revenue collection from forestry operations. According to interview respondents, however, this claim to self-sustenance was knowingly false on the part of the reformers, in that they knew there was 'no way' it would be sustainable, or less perversely that it was only 'on its way to being sustainable'.[64] The claim was justified in their eyes, in that it was in line with their vision for forestry, where the building of a viable timber industry (which had been in decline after a lack of planting in the Amin years) was seen as crucial to the revival of the sector and the protection of 'natural forests'.[65] As Obua et al contend, 'Little attention had been paid to development of commercial forests which should have provided alternative forest products and services to relieve the pressure on natural forests and conserve biodiversity'.[66] However, the result has seen a form of de-territorialisation (see below), with the faltering decentralisation of the DFS to local government after substantive funding for it failed to materialise,[67] and a lack of oversight, reporting and coordination facility from the FSSD, which did not receive sufficient resources or have sufficient capacity to fulfil its mandate.[68]

Outscaling

Outscaling involves the turn to the private sector, and to calculative practices, as an attempt to address both revenue shortfalls from protected areas and central forest reserves, and shortfalls in state funding for forestry conservation, both of which compromise the perceived 'economic sustainability of conservation policy'.[69] Here extra-state, civil society or market-based configurations such as plantation forestry become increasingly represented in managerial plans, regulation and governance of conservation and protected areas.[70] The forestry reforms institutionalised a funding focus on facilitating private timber planting on CFRs, on the one hand, and the deregulation of the control and use of non-CFR 'forests' to 'private forest owners', on the other. Whereas previously people had perceived such forests outside of protected areas as 'government forests',[71] which UFD staff had nominally regulated, under the NFTPA 2003 'private' forests were codified as being on 'private land' (even though the category does not exist elsewhere in Uganda). The idea was that, with rationalisation (and tree tenure security), 'communities' could then better utilise their land, or sell, mortgage or lease its assets to corporations or other individuals who could use it efficiently.[72]

Similarly over time a greater reliance on market environmentalism encouraged the 'responsibilisation' and financialisation of individuals and companies as neoliberal 'carbon fixing subjects', seen as effective first responders to climate and deforestation crises in Uganda. With respect to the private sector the framing of a deforestation crisis, represented by the perceived need to establish 150,000 ha of timber plantations by 2020 to address the 'shortfall' in demand,[73] resulted in the encouragement of private planting to augment the low capacity of government actors. This involved the establishment of growers and marketing associations, new biological and chemical regimes of production, and the attraction and provision of incentives such as tax cuts for the 'Big Four' plantation companies: New Forests (UK), Green Resources AS (Norway), Global Woods (Germany) and Nile Ply (Uganda), all with the investment capacity required to advance the vision for forestry. As a retired academic from Makerere University (active in environmental policy consultations in the 1980s and 1990s) put it:

> Before, we had biodiversity. Now there is sustainability, and the investigation of ways to harness biodiversity for production. So the forest of today is a plantation. If you look at the current trend people are looking at natural reserves as areas secluded from development. In the 90s the elite were the vanguard of protection, but today it is seen as negative land, negative space. It is 'degraded' to be used for other purposes.[74]

The overall direction in policy that resulted was a significant reorientation towards market based activities – in both plantation forestry and carbon forestry -that reimagines forestry as a form of commercial agriculture within the bio-economy, reducing complex ecosystems to biomass (including carbon and timber) divorced from state conservation forestry models.[75] Yet plantation targets have not been reached, and Jagger disputes the claim that the reforms improved livelihoods, claiming that on private forest land overseen by the decentralised DFS there had been no significant change in average annual household income from forests, and that for households adjacent to forest reserves increases were limited to households in the highest income quartile (and were primarily attributed to the sale of illegally harvested timber).[76]

Downscaling

I have already discussed ostensible forms of downscaling through decentralisation, but there is a variety of examples of new calculative governance structures which 'reach' down to the local level, and in so doing re-territorialise forestry governance. The inception of Forest Stewardship Council (FSC) accreditation in Uganda to promote 'sustainable timber exports', and the genesis of carbon forestry projects are the manifestation of this form of self-regulated governance. The implications are that the modality of governance shifts from the control of territories (established during territorialisation in Uganda), to the control of flows of commodities (such as FSC-certified timber and the carbon in carbon forestry).[77]

The Forest Functions and Classification document produced by the NFA in 2005, after an inventory process which attempted the 'zoning' of forest reserves for specific purposes and according to specific criteria, is another. Perhaps the most interesting, however, was the USAID[78] Uganda Mission's Strategic Criteria for Rural Investments in Productivity (SCRIP) programme (which ran from 2001 to 2006), which exemplifies the new modality of governance, or set of technologies of power and calculation,[79] mobilised in relation to the state (and in the interests of capital), with the express aim of 'improv[ing] policy and investment' in land use. This intervention essentially sets out economic valuations of 'natural capital' by identifying goods and services as well as the loss of biodiversity in all ecosystems in monetary terms.[80] It estimates carbon and biodiversity values, noting than forests, despite making up only 4% of the land, contribute 18% to Uganda's total Biodiversity Values (estimated at 750 US Dollars a year for all Uganda's 'terrestrial ecosystems', or around US$200 per hectare at '2002 values'). It also ranks National Parks, Key Forests and other areas (including wetlands and savannahs) according to value criteria, including the 'risks' of conflicts in land use, and trade-offs, essentially between biodiversity and human use that arise, identifying areas suitable for 'better' use of natural resources.

Such calculations are of utility in two ways, but it is also evident that attempts at re-territorialisation to promote the downscaling of governance are limited,[81] particularly in terms of biodiversity conservation. The forest inventory programme up to 2005, for instance, only surveyed 65 of some 700 CFRs and, furthermore, those which were seen as suitable for zoning for productive and conservation uses, respectively. The first respect in which new calculative frameworks are of utility is for facilitating private planting, with the Small Production Grant Scheme (SPGS) for instance making use of cherry-picked CFRs from the exercises, which isolate territories suitable for investment. A second element of a flow-based governance involves the calculation of carbon stored in woody biomass and commodified as emission reductions units, and a risk calculus around climate change mitigation. Here carbon finance, stemming from flows of carbon, is framed as a new source of funding for development and conservation in the global South, and a rapidly growing business opportunity for those who develop and broker projects and credits.

Instituting carbon forestry involves the responsibilisation and financialisation of individuals, companies and NGOs variously as carbon credit developers, purchasers and carbon fixers. Here governance comes to function not around actual landscapes, but around their reduction to and abstract representation in virtual, calculative spaces and parameters (such as timber planting targets and projected emissions reductions units) to make a virtual matrix that renders individuals able to 'self-regulate' and perform governance roles in relation, in this case, to deforestation, logging practice and carbon emissions. The ideology which underpins this is a neoliberal environmentality, distinct from a sovereign environmentality – the disciplinary form of control over territory by the state.[82]

The resulting carbon forestry interventions themselves include REDD+, Afforestation/reforestation Clean Development Mechanism (A/R CDM) projects and Voluntary Carbon Market (VCM) projects, and comprise one facet of what can be termed market environmentalism: the extension of market logics to the governance of the environment. There were nine projects in Uganda engaged in this research: three prospective REDD projects, three A/R CDM initiatives and three VCM projects. Six of these projects described above are on protected areas, and four on so-called private lands.[83] There have been controversies surrounding many of these projects,[84] and despite critiques and opposition to carbon forestry,[85] advocates argue that, when funding finally arrives, through a process of learning by doing and adjustments to market-oriented incentive structures, interventions can improve.

The new hybrid governance form: de-territorialisation and counter-performativities

In describing changes towards a neoliberal flow-based governance form, and a re-territorialisation of forest resources, I do not claim that that the older form of territorial control has been completely superseded. To the contrary, a hybrid form arises, in which *de jure* state territoriality over protected forest areas in particular remains unquestioned, but is undercut, first, by the new flow-based form of governance and, second, in the ways in which the everyday territoriality of the forest estate is inherently contested. Neither are re-territorialisations and de-territorialisation mutually exclusive processes in the context I describe them, but two sides of the same coin. The concept of de-territorialisation is useful when considering processes of entropy linked to the neoliberalisation of forestry governance – in which state control of forestry resources, as well as the territoriality of protected forest areas, is unevenly weakened by the inclination towards flow-based arrangements. In this light government support for forestry has been steadily declining. According to the National Development Plan for 2010–15, the forest sector will have 96% of its planned budget not funded, while donor funding has shifted away from the funding of what were perceived as corrupt institutions towards carbon forestry activities such as Uganda's Ready for REDD process. Under the old Uganda Forest Department there were cross-subsidies maintaining the costs and management of the entire forest estate, whereby staff oversaw not only CFRs but what was described as 'natural forest' outside of Crown lands (on differing land tenure types in Uganda).[86] After the reforms the NFA garnered the lion's share of resources, while the decentralised DFS that emerged was weak and underfunded, with a diminished and decentralised regulatory capacity, despite significant responsibilities. According to Turyahabwe, the author of a 2006 review of forest decentralisation, the cumulative result of the new dispensation was that 'the DFS never took off, it just remained on paper'.[87] The NFA is weak, with capacity shortages and a lack of resources, after it failed to attain its self-professed self-sustainability. This is particularly so at remote CFR locations. Because of data and capacity limitations, surveys across the whole forest protected estate are impossible; in many senses the institutions are working blind, and regularly have problems 'opening the boundaries' (an idiom used to describe protected area demarcation and defence) of the reserves.

In this context deforestation in forests has increased, through so-called 'encroachment' (informal or illegal access) into natural forests, where tree cutting for timber and charcoal production are prominent, and where land conversion for agriculture and settlement has occurred in forests on 'private lands'[88]. These pressures for encroachment have intensified with increasing population pressure, land scarcity and growing landlessness exacerbated by agricultural intensification (under Uganda's Plan for Modernization of Agriculture (PMA) and large-scale land acquisitions).[89] Without effective institutions to mitigate or manage these pressures, evictions and marginalisation of encroachers (whose rights as citizens are suspended in a state of exception) and local communities have accompanied plantation establishment and carbon forestry project implementation,[90] re-enforcing, often through violence, continuously contested, 'hard edged' boundaries of protected areas. Such activities were most noticeable during the donor-funded FRMCP. In such circumstances the most affected are 'the poor and vulnerable who lack financial and other resources to enable them to challenge the evictions or enable them to acquire alternative settlement...some of whom were born on the reserves where they have been living for decades, and do not know of any other home'.[91] In another paper I term such processes forms of 'social sacrifice' that accompany market environmentalism.[92]

Feedback effects have also intensified the problems affecting forestry governance, in particular stemming from the reorganised inter-relationships between the NFA, processes of 'donorisation' and the NRM-dominated state after the neoliberalisation of the forestry sector. The external support lent to the NRM regime in general – and to the forestry sector in particular – through a process of 'donorisation' has indirectly supported regime consolidation, patronage politics, the 'informalisation' of the state, as well as the perpetuation of illegal activities connected to the maintenance of power.[93] Paradoxically, then, strong state capacity in Uganda may exacerbate processes of deforestation and illegal logging and encroachment, through informal agency of forestry actors embedded within the state at both local and national scales.[94]

As the head of the chronically underfunded oversight body the FSSD put it, 'The NFA was funded to do their work; but they are only 15% of the estate, and when the regulator wasn't there to see their contracts, they have done it the way they feel...and corruption has been growing fast'.[95] This increase in corruption, and the public scandals that dogged the institute in 2009 that highlighted it, followed the breakdown of NFA capacity, and the resignation of its board after confrontation with the central NRM government. By this stage the quasi-privatised and donor-initiated parastatal NFA was in a weak position, after failing its mandate of 'self-sufficiency'. Simultaneously and paradoxically, however, it could be interpreted as having become 'too strong', when it resisted presidential orders for the de-gazetting of a number of the CFRs for industrial agriculture, including palm oil at the Kalangala Islands and sugarcane at the

Mabira Central Forest Reserve.[96] This led to pressure for the resignation of donor-approved appointees, and pressure on remaining technocrats to 'work to the central government's agenda – which was tied to further state funding.[97]

In the context described, 'political interference' in the maintenance of the protected forest estate has increased,[98] and a carefully detailed report by the Uganda Anti-Corruption Coalition (ACCU) articles both cases of 'forest give away' to those politicians mandated to manage the forest sector 'in trust' for the people of Uganda. Following an official, but in reality selective, presidential ban on the licensing of allocations of CFR land in 2006 – ostensibly because the NFA was 'doing it badly' – patronage came to play a central role in the allocation of concessions and leases of CFR territory.[99] Furthermore, the influence of informal or illegal networks has become more prevalent after the forest sector reform, nested within a trend which has seen the increase of illegal activities in the forestry sector since 2000.[100] In individual forest territories informal access regimes are lubricated by payments where officials turn a blind eye, or seek to make the best of a bad situation as some see it.[101] In another example, after a ban by the Minister of Water and Environment on timber cutting to address illegalities in 2012, the targeted dealers went to parliament and were given an audience with the Natural Resources committee and the ban was soon dropped.

While it would be tempting to dismiss these cases as the mere moral failings of politicians and officials (and this certainly does play a role), such forms of de-territorialisation have as much to do with the complexities of the new hybrid governance arrangement, and with the problems of maintaining an exclusionary management regime, both practically and politically. At the national state the cessation of a massive boundary reassertion, and the eviction campaign that was instituted against encroachers after the establishment of the NFA,[102] evidences such complexities. The lobbying of local civil society about the welfare of the vulnerable certainly contributed to the cessation, but it was *Realpolitik* that underpinned it – it does not make sense to alienate potential voters by evicting them, especially in the context of declining regime popularity.

Conclusion

This paper has expanded post-structuralist lines of enquiry in assemblage and territory into the sphere of market environmentalism, exploring how it localises in a particular place, Uganda. It traced three processes of territorialisation (drawing from classical accounts of state territorialisation of forest resources), re-territorialisation and de-territorialisation, exploring Deleuzo-Guattarian formulations of territorialised assemblage, to examine the relationships and reformulations of power, knowledge and social space that come together within contemporary neoliberal forestry governance. I have argued that the control of communities of trees and humans is changing, through processes of re-territorialisation, from the exclusive control of territories to a hybrid governance with leanings towards control through flows of both carbon credits and biomass. These changes privilege non-state actors, and entail negative externalities and substantive de-territorialisations of state territory.

What is evident is that carbon forestry is not only an attempt at the capture and long-term storage of atmospheric carbon in individual project locations but a political-economic technology to render climate change governable and to sequester governance-beyond-the state. It should be seen as a calculative mechanism, directed through a neoliberal environmentality, for controlling actors engaged in the rural under the organising principle of carbon. Here the state increasingly interacts with non-state actors, through the exercise of state reach, to legitimise its authority and insulate itself from criticism, rather than to mean-ingfully decentralise control. In looking forward towards a more progressive alignment, the analysis here 'de-territorialises' the symbolic edifice of the forest estate and managerialist state policy, and hints that we should be concerned with 'sociabilities' and 'communities' and not merely the capture of value and utilisa-tion of resources. As such it may be necessary to revisit the territoriality of the forest estate itself and the uneven power relationships that underpin it.

Acknowledgements

I would like to acknowledge the organisers of the 'Green Economy in the South' conference, for which this paper was originally submitted, and the University of Kwazulu-Natal's Department of Geography Cartographer, Brice Gijsbertsen.

Notes

1. Bridge, "Resource Geographies I"; Castree, "Neoliberalising Nature"; and Bakker, "Neoliberalizing Nature?"
2. Larner, "C-Change?"
3. Leach and Mearns, "Environmental Change and Policy."
4. Murray Li, *The Will to Improve*.
5. Mwangi and Wardell, "Multi-level Governance of Forest Resources."
6. Jonas and Bridge, "Governing Nature," 3.
7. Murray Li, "Practices of Assemblage and Community Forest Management"; and Descheneau and Patersen, "Between Desire and Routine."
8. Smith, *Uneven Development*; and Vandergeest and Peluso, "Territorialisation and State Power in Thailand."
9. Benterrak et al., *Reading the Country*.
10. See Vaccaro et al., "Political Ecology and Conservation Policies."
11. See Allen and Cochrane, "Assemblages of State Power."
12. Leach and Scoones, *Carbon Conflicts and Forest Landscapes*; and Nel, *Assembling Value in Carbon Forestry*.
13. Deleuze and Guattari, *A Thousand Plateaus*.
14. Delaney, *Territory*, 205.
15. Delaney, "Territory and Territoriality," 207.
16. Nel, *Assembling Value in Carbon Forestry*.
17. Anderson and McFarlane, "Assemblage and Geography," 149.
18. Lang and Byakola, *A Funny Place to Store Carbon*; and Twongyirwe et al., "REDD+ at Crossroads?"

19. Turyahabwe and Banana, "An Overview."
20. Leach and Scoones, *Carbon Conflicts and Forest Landscapes*.
21. Mamdani, *Citizen and Subject*, contends that what we call Uganda is not that 'thing' that emerged as the result of an inevitable, linear process of colonisation, but a bundle of disparate, varying and uneven tensions that have been temporarily and unsatisfactorily resolved through violence.
22. Swyngedouw, "Governance Innovation and the Citizen."
23. Sassen, "Neither Global nor National."
24. Lemke, "Foucault, Governmentality, and Critique," 50.
25. Oels, "Rendering Climate Change Governable."
26. Delaney, *Territory*, 205.
27. NFA, *History of Forest Inventory*.
28. Mamdani, *Citizen and Subject*, 7–10.
29. Cavanagh and Himmelfarb, "'Much in Blood and Money'."
30. Mamdani, *The Contemporary Ugandan Discourse on Customary Tenure*.
31. Kjekshus, *Ecological Control and Economic Development*.
32. Doyle, "An Environmental History of the Kingdom of Bunyoro."
33. Vandergeest and Peluso, "Territorialisation and State Power in Thailand."
34. Scott, *Seeing like a State*.
35. For another example of such 'political forestry', or 'national forests', see Peluso and Vandergeest, "Genealogies of the Political Forest."
36. Ibid.
37. Peluso and Lund, *New Frontiers of Land Control*, 647.
38. Turyahabwe and Banana, "An Overview."
39. Ibid., 2.
40. Neumann, "Nature–State–Territory."
41. Cowie, "Preserve or Destroy?"; and Struhsaker, "Forest and Primate Conservation."
42. Langdale Brown, *Biomass Vegetation of Uganda*.
43. Webster and Osmaston, *A History of the Uganda Forest Department*.
44. Ibid., vii.
45. Nel and Hill, "Constructing Walls of Carbon"; and Cavanagh and Himmelfarb, "'Much in Blood and Money'."
46. Adams, *Against Extinction*.
47. Nsita, "Decentralisation and Forest Management."
48. See Webster and Osmaston, *A History of the Uganda Forest Department*.
49. Nsita, "Decentralisation and Forest Management," 1.
50. Guweddeko, *Anatomy of Museveni and Mengo Crisis*.
51. Van de Wiel, "Uganda."
52. Marquardt, "Settlement and Resettlement", describes how ongoing encroachment on to *de jure* protected forest areas has continued over decades: first to maintain or reclaim land which was lost in the creation of reserves; second to relocate into vacant, seemingly unclaimed land; and third to expand land from areas outside the reserves, encouraged by the breakdown of central government authority, the inability to control population movement and official incentives to alleviate land pressures in surrounding district areas.
53. Swyngedouw, "Governance Innovation and the Citizen."
54. Jessop, "Hollowing out the Nation-state."
55. Cohen and Bakker, "The Eco-scalar Fix," 2.
56. Jonas and Bridge, "Governing Nature," 3.
57. Duffield, "Governing the Borderlands."
58. Comment made at the REDD Steering Committee Meeting, Kampala, August 2012.
59. Allen and Cochrane, "Assemblages of State Power," 1074.
60. Jacovelli, "Uganda's Sawlog Production Grant Scheme."
61. Nsita, "Facing the Challenges of Change"; and Jagger, *Forest Incomes*.
62. Khisa, "The Making of the 'Informal State'."
63. Interview, former Forestry Actor, November 2012.
64. Ibid.
65. Obua et al., "Status of Forests in Uganda," 853.
66. District structures were undergoing a similar restructuring at the time, involving the decentralisation of other state functions to local government level. This provided an opportunity to 'tack on' a structure and a small staffing complement of two District Forest Officers per district to constitute the DFS.
67. FSSD Official, Kampala, October 2012.
68. Corson, "Territorialization, Enclosure and Neoliberalism."
69. Büscher et al., "Towards a Synthesized Critique."
70. Community Focus group responses, Hoima and Masindi, September 2012.
71. For a parallel in Indonesia, see Murray Li, *The Will to Improve*, 285.
72. Unique, *UNIQUE report for SPGS*.

73. Interview with Paul Nusali, Kampala, August 2012.
74. Interview with Paul Jacovelli, Kampala, November 2012.
75. Jagger, *Forest Incomes*.
76. Sikor et al., "Global Land Governance."
77. USAID. *Intergrated Stretegic Plan for USAID's Program in Uganda.*
78. Dean, *Governmentality*.
79. Pomeroy et al., *Uganda Ecosystem*.
80. For instance, a study by Makerere University from 2008 to 2011 of 35 CFRs in Eastern Uganda found that most had not been zoned, ostensibly in order to protect valuable and 'rare' biodiversity as mandated in the Nature Conservation Master Plan of 2002 (New Vision 2012). Few comprehensive 'forest' inventories were done by the debilitated NFA after 1999, because of their relative cost and time consuming nature.
81. Fletcher, "Neoliberal Environmentality."
82. See Nel and Hill, "Comparing Project Orientations and Commercialisation Logics."
83. Grainger and Geary, *The New Forests Company*; Cavanagh and Benjaminson, "Virtual Nature, Violent Accumulation"; Nel and Hill, "Constructing Walls of Carbon"; and Lyons and Westoby, "Carbon Colonialism."
84. For example, by groups such as the No-REDD+ in Africa Network. See also Lohman, *Neoliberalism and the Calculable World*; and Bond et al., *The CDM cannot Deliver*.
85. The four tenure types in Uganda include Mailo (a traditional form of ownership specific to the Buganda state), Freehold, Leasehold and Communal tenure.
86. Interview, Makerere University, August 2012.
87. Interview, NFA Director of Natural Forests, Kampala, October 2013.
88. Martinello, "The Accumulation of Dispossession."
89. Lyons and Westoby, "Carbon Colonialism"; and Cavanagh and Benjaminson, "Virtual Nature, Violent Accumulation."
90. Mugyenyi et al., *Balancing Nature Conservation and Livelihoods*, 5.
91. Nel, "The Choreography of Sacrifice."
92. See Khisa, "The Making of the 'Informal State'," who discusses 'informalisation' as a distinct mode of organising and broadcasting power that simultaneously centralises and fragments the state system. See also Mwenda and Tumushabe, *A Political Economy Analysis*; and Mwenda and Tangri, "Patronage Politics."
93. Cavanagh and Benjaminsen, "Guerrilla Agriculture?"; and Cavanagh et al., "Securitizing REDD+?"
94. Interview, October 2012.
95. Carmody, "It's easy to rule a Poor Man."
96. Interviews with ex-UFD and FRMCP secretariat members, Nakulaabye, Kampala, October 2012.
97. Such interference includes the directive of the president in 2010 in Mount Elgon siding with the community against the Uganda Wildlife Authority, and promises by politicians to encroachers that they would de-gazette areas if they were voted for.
98. Interview, former NFA official, Kampala, July 2012.
99. Anti-Corruption Coalition Uganda, *Namanve Forest Report*.
100. Interview, NFA range manager, Kiboga, October 2012.
101. Mugyenyi et al., *Balancing Nature Conservation and Livelihoods*.

Bibliography

Adams, W. M. *Against Extinction: The Story of Conservation*. London: Earthscan, 2013.

Anderson, B., and C. McFarlane. "Assemblage and Geography." *Area* 43, no. 2 (2011): 124–127.

Anti-Corruption Coalition Uganda. *Namanve Forest Report: Environmental Crisis looms as Forests come under Threat: Cases of Forest Giveaway and Illegal Activity*. Kampala: Anti-Corruption Coalition Uganda, 2010.

Allen, J., and A. Cochrane. "Assemblages of State Power: Topological Shifts in the Organization of Government and Politics." *Antipode* 42, no. 5 (2010): 1071–1089.

Bakker, K. "Neoliberalizing Nature? Market Environmentalism in Water Supply in England and Wales." *Annals of the Association of American Geographers* 95, no. 3 (2005): 542–565.

Benterrak, K., S. Muecke, and P. Roe. *Reading the Country: Introduction to Nomadology*. Fremantle: Fremantle Arts Centre Press, 1984. http://www.getcited.org/pub/102482313.

Bond, P., K. Sharife, F. Allen, B. Amisi, K. Brunner, R. Castel-Branco, D. Dorsey, G. Gambirazzio, T. Hathaway, and A. Nel. *The CDM cannot deliver the Money to Africa*. Durban: Centre for Civil Society, 2012. http://www.ejolt.org/wordpress/wp-content/uploads/2013/01/121221_EJOLT_2_Low.pdf.

Bridge, G. "Resource Geographies I: Making Carbon Economies, Old and New." *Progress in Human Geography* 35, no. 6 (2011): 820–834.

Büscher, B., S. Sullivan, K. Neves, J. Igoe, and D. Brockington. "Towards a Synthesized Critique of Neoliberal Biodiversity Conservation." *Capitalism Nature Socialism* 23, no. 2 (2012): 4–30.

Castree, N. "Neoliberalising Nature: The Logics of Deregulation and Reregulation." *Environment and Planning A* 40, no. 1 (2008): 131–153.

Carmody, P. "It's easy to rule a Poor Man." Journal manuscript, Trinity College Dublin.

Cavanagh, C. J., and T. A. Benjaminsen. "Guerrilla Agriculture? A Biopolitical Guide to Illicit Cultivation within an IUCN Category II Protected Area." *Journal of Peasant Studies* 42, nos. 3–4 (2015): 725–745.

Cavanagh, J. C., and T. A. Benjaminsen. "Virtual Nature, Violent Accumulation: A Critical Political Ecology of Carbon Market Failure at Mt. Elgon, Uganda." *Geoforum* 56 (2014): 55–65.

Cavanagh, J. C., and D. Himmelfarb. "'Much in Blood and Money': Colonial Conservation, Resistance, and the State on the Margins of the Uganda Protectorate, 1900–1962." *Antipode* 47, no. 1 (2015): 55–73.

Cavanagh, C. J., P. O. Vedeld, and L. T. Trædal. "Securitizing REDD+? Problematizing the Emerging Illegal Timber Trade and Forest Carbon Interface in East Africa." *Geoforum* 60 (2015): 72–82.

Cohen, A., and K. Bakker. "The Eco-scalar Fix: Rescaling Environmental Governance and the Politics of Ecological Boundaries in Alberta, Canada." *Environment and Planning D: Society and Space* 32, no. 1 (2014): 128–146.

Corson, C. "Territorialization, Enclosure and Neoliberalism: Non-state Influence in Struggles over Madagascar's Forests." *Journal of Peasant Studies* 38, no. 4 (2011): 703–726.

Cowie, M. "Preserve or Destroy?" *Oryx* 3, no. 1 (1955): 9–11.

Dean, M. *Governmentality: Power and Rule in Modern Society.* London: Sage Publications, 2009.

Descheneau, P., and M. Paterson. "Between Desire and Routine: Assembling Environment and Finance in Carbon Markets." *Antipode* 43, no. 3 (2011): 662–681.

Delaney, D. *Territory: A Short Introduction.* London: Wiley, 2008.

Delaney, D. "Territory and Territoriality." In *International Encyclopedia of Human Geography*, edited by R. Kitchin and N. Thrift, 196–208. Amsterdam: Elsevier, 2009.

Deleuze, G., and P. F. Guattari. *A Thousand Plateaus: Capitalism and Schizophrenia.* Minneapolis, MN: University of Minnesota Press, 1987.

Doyle, S. *An Environmental History of the Kingdom of Bunyoro in Western Uganda, from c.1860 to 1940.* PhD diss.: University of Cambridge, 1998.

Duffield, M. "Governing the Borderlands: Decoding the Power of Aid." *Disasters* 25, no. 4 (2001): 308–320.

Fletcher, R. "Neoliberal Environmentality: Towards a Poststructuralist Political Ecology of the Conservation Debate." *Conservation and Society* 8, no. 3 (2010): 171–181.

Grainger, M., and K. Geary. *The New Forests Company and its Uganda Plantations.* Washington, DC: Oxfam International, 2011.

Guweddeko, F. *Anatomy of Museveni and Mengo Crisis – 1806.* Uganda: Independent, 2009. http://www.independent.co.ug/index.php/component/content/article/1806?format=pdf.

Jacovelli, P. "Uganda's Sawlog Production Grant Scheme: A Success Story from Africa." *International Forestry Review* 11, no. 1 (2009): 119–125.

Jagger, P. *Forest Incomes after Uganda's Forest Sector Reforms.* Washington, DC: CIGAR, 2008.

Jessop, B. "Hollowing out the Nation-state and Multilevel Governance." In *A Handbook of Comparative Social Policy*, edited by P. Kennet, 11–26. Cheltenham: Edward Elgar, 2004.

Jonas, A. E. G., and G. Bridge. "Governing Nature: The Re-regulation of Resources, Land-use Planning, and Nature Conservation." *Social Science Quarterly* 84, no. 4 (2003): 958–962.

Khisa, M. "The Making of the 'Informal State' in Uganda." *Africa Development* 38, nos. 1–2 (2013): 191–226.

Kjekshus, H. *Ecological Control and Economic Development in East African History.* London: Heinemann, 1977.

Lang, C., and T. Byakola. *A Funny Place to Store Carbon: UWA-FACE Foundation's Tree Planting Project in Mount Elgon National Park, Uganda.* Montevideo & Moreton: World Rainforest Movement, 2006. http://wrm.org.uy/wp-content/uploads/2013/02/Place_Store_Carbon.pdf.

Langdale-Brown, I. *Biomass Vegetation of Uganda.* 6 vols. Kampala: Kawanda Research Station, 1960.

Larner, W. "C-Change? Geographies of Crisis." *Dialogues in Human Geography* 1, no. 3 (2011): 319–335.

Leach, M., and R. Mearns. "Environmental Change and Policy." In *The Lie of the Land: Challenging Received Wisdom on the African Environment*, edited by M Leach and R Mearns. Oxford: James Currey, 1996.

Leach, M., and I. Scoones (eds.). *Carbon Conflicts and Forest Landscapes in Africa.* Abingdon: Routledge, 2015.

Lemke, T. "Foucault, Governmentality, and Critique." *Rethinking Marxism* 14, no. 3 (2002): 49–64.

Lohman, L. *Neoliberalism and the Calculable World: The Rise of Carbon Trading.* London: Zed Books, 2009.

Lyons, K., and P. Westoby. "Carbon Colonialism and the New Land Grab." *Journal of Rural Studies* 36, no. 3 (2014): 13–21.

Mamdani, M. *Citizen and Subject: Contemporary Africa and the Legacy of Late Colonialism.* Princeton, NJ: Princeton University Press, 1996.

Mamdani, M. The Contemporary Ugandan Discourse on Customary Tenure: Some *Theoretical Considerations.* Kampala: Makerere Institute of Social Research, 2013.

Marquardt, M. "Settlement and Resettlement: Experience from Uganda's National Parks and Reserves." In *Involuntary Resettlement in Africa: Selected Papers from a Conference on Environment and Settlement Issues in Africa*, edited by C. C. Cook. Washington, DC: World Bank Publications, 1994.

Martinello, G. "The Accumulation of Dispossession, Agrarian Change and Resistance in northern Uganda." Manuscript, Makerere Institute for Social and Economic Research.

Mugyenyi, O., B. Twesigye, and E. Muhereza. *Balancing Nature Conservation and Livelihoods: A Legal Analysis of the Forestry Evictions by the National Forestry Authority*. ACODE Policy Briefing Series 13. Kampala: ACODE, 2005.

Murray Li, T. *The Will to Improve: Governmentality, Development, and the Practice of Politics*. Durham, NC: Duke University Press, 2007.

Li, Murray. "T. "Practices of Assemblage and Community Forest Management"." *Economy and Society* 36, no. 2 (2007): 263–293.

Mwangi, E., and A. Wardell. "Multi-level Governance of Forest Resources." *International Journal of the Commons* 6, no. 2 (2012): 79–103.

Mwenda, A., and R. Tangri. "Patronage Politics, Donor Reforms, and Regime Consolidation in Uganda." *African Affairs* 104, no. 416 (2005): 449–467.

Mwenda, A., and G. W. Tumushabe. *A Political Economy Analysis of the Environmental and Natural Resources Sector in Uganda*. Kampala: World Bank, 2011.

Nel, A. *Assembling Value in Carbon Forestry: Practices of Assemblage, Overflows and Counter-performativities in Ugandan Carbon Forestry*. Working Paper Series 10. Leverhulme Centre for the Study of Value, Manchester University, 2015.

Nel, A. "The Choreography of Sacrifice: Neoliberal Biopolitics, Environmental Damage and Assemblages of Market Environmentalism." Manuscript submitted to *Geoforum*.

Nel, A., and D. Hill. "Comparing Project Orientations and Commercialisation Logics within Carbon Forestry Projects in Eastern and Southern Africa." *Capitalism Nature Socialism* 25, no. 4 (2014): 19–35.

Nel, A., and D. Hill. "Constructing Walls of Carbon – The Complexities of Community, Carbon Sequestration and Protected Areas in Uganda." *Journal of Contemporary African Studies* 31, no. 3 (2013): 421–440.

Neumann, R. "Nature–State–Territory: Toward a Critical Theorization of Conservation Enclosures." In *Liberation Ecologies: Environment, Development, Social Movements*, edited by M. Watts and R. Peet. London: Routledge, 2004.

Nsita, S. "Facing the Challenges of Change: The Uganda Sector Reforms 2001–2011." In Forest Governance Learning Group Symposium on Overcoming Challenges in Forest Governance after a Decade of Reform. Kampala, 2012.

Nsita, S. "Decentralisation and Forest Management in Uganda in Support of the Intercessional Country-led Initiative on Decentralisation." Federal Systems of Forestry and National Forestry Programmes, 2002.

NFA. *History of Forest Inventory: Forest Inventory Methods in Use at NFA*. National Forestry Authority Supplement, Kampala, 2009.

Obua, J., J. G. Agea, and J. J. Ogwal. "Status of Forests in Uganda." *African Journal of Ecology* 48, no. 4 (2010): 853–859.

Oels, A. "Rendering Climate Change Governable: From Biopower to Advanced Liberal Government?" *Journal of Environmental Policy & Planning* 7, no. 3 (2005): 185–207.

Peluso, N., and C. Lund. *New Frontiers of Land Control*. London: Taylor & Francis Group, 2012.

Peluso, N. L., and P. Vandergeest. "Genealogies of the Political Forest and Customary Rights in Indonesia, Malaysia, and Thailand." *Journal of Asian Studies* 60, no. 3 (2001): 761–812.

Pomeroy, D., H. Tushabe, P. Mwima, and P. Kasoma. *Uganda Ecosystem and Protected Area Characterisation*. Kampala: Makerere University Institute of Environment and Natural Resources (MUIENR), 2002.

Sassen, S. "Neither Global nor National: Novel Assemblages of Territory, Authority and Rights." *Ethics & Global Politics* 1, nos. 1–2(2008): 61–79.

Scott, James C. *Seeing like a state: How certain schemes to improve the human condition have failed*. New Haven: Yale University Press, 1998.

Sikor, T., G. Auld, A. J. Bebbington, T. A. Benjaminsen, B. S. Gentry, C. Hunsberger, A-M. Izac, et al. "Global Land Governance: From Territory to Flow?" *Current Opinion in Environmental Sustainability* 5, no. 5 (2013): 522–527.

Smith, N. *Uneven Development: Nature, Capital, and the Production of Space*. Athens: University of Georgia Press, 2008.

Struhsaker, T. "Forest and Primate Conservation in East Africa." *African Journal of Ecology* 19, nos. 1–2 (1981): 99–114.

Swyngedouw, E. "Governance Innovation and the Citizen: The Janus Face of Governance-beyond-the-state." *Urban Studies* 43, no. 11 (2005): 1991–2006.

Turyahabwe, N., and A. Banana. "An Overview of History and Development of Forest Policy and Legislation in Uganda." *International Forestry Review* 10, no. 4 (2008): 641–656.

Twongyirwe, R., D. Sheil, C. Sandbrook, and L. C. Sandbrook. "REDD+ at Crossroads? The Opportunities and Challenges of REDD+ for Conservation and Human Welfare in South West Uganda." 2014. http://www.repository.cam.ac.uk/handle/1810/246312.

Unique. *UNIQUE Report for SPGS on Projected Timber Demand by 2020*. Kampala: Unique Forestry and Land Use, 2010.

USAID. *Intergrated Stretegic Plan for USAID's Program in Uganda*. Kampala: USAID, 2001.

Vaccaro, I., O. Beltran, and A. Paquet. "Political Ecology and Conservation Policies: Some Theoretical Genealogies." *Journal of Political Ecology* 20 (2013): 255–272.

Van de Wiel, A. "Uganda: The State and the Nation." *Pambazuka News*, 541 (2011). http://www.pambazuka.org/en/category/features/75207.

Vandergeest, P., and N. L. Peluso. "Territorialisation and State Power in Thailand." *Theory and Society* 24, no. 3 (1995): 385–426.

Webster, G., and H. Osmaston. *A History of the Uganda Forest Department 1951–1965*. London: Commonwealth Secretariat, 2003.

Inverting the moral economy: the case of land acquisitions for forest plantations in Tanzania

M.F. Olwig[a], C. Noe[b], R. Kangalawe[c] and E. Luoga[d]

[a]Department of Society and Globalisation, Roskilde University, Denmark; [b]Geography Department, University of Dar es Salaam, Tanzania; [c]Institute of Resource Assessment, University of Dar es Salaam, Tanzania; [d]Faculty of Forest Mensuration, Sokoine University of Agriculture, Morogoro, Tanzania

Governments, donors and investors often promote land acquisitions for forest plantations as global climate change mitigation via carbon sequestration. Investors' forestry thereby becomes part of a global moral economy imaginary. Using examples from Tanzania we critically examine the global moral economy's narrative foundation, which presents trees as axiomatically 'green', 'idle' land as waste and economic investments as benefiting the relevant communities. In this way the traditional supposition of the moral economy as invoked by the economic underclass to maintain the basis of their subsistence is inverted and subverted, at a potentially serious cost to the subjects of such land acquisition.

Introduction

The expansion of land acquisitions in Africa is in part motivated and justified by a concern with the need to mitigate global climate change. Tree planting, which plays 'a considerable role in terms of approved land deals and planted area',[1] is thus promoted as a climate change mitigation measure. Forestry plantations thereby enable investors to combine incomes from wood production with incomes from carbon sequestration.[2] These plantations of usually non-native tree species are being established on what investors and local governments often refer to as idle or under-utilised land that has not necessarily been forested before.

This paper argues that, when investors link forestation with carbon sequestration, their economic activities are framed as supporting a global environmental common good by which the establishment of forestry plantations becomes part of a global moral economy imaginary. Using illustrations from a study of two

villages in Mufindi District, Tanzania the paper critically examines the foundation of this narrative – or what could be called the 'common wisdoms'[3] – of this
global moral economy imaginary. The narrative foundation, we find, is comprised of three key 'storylines', namely that trees are axiomatically 'green,'
'idle' land is waste and global investments lead to local benefits. We further
suggest that, while investment in forestation for profit is legitimised by the
notion of a global moral economy, the traditional supposition of the moral economy as invoked by the economic underclass to maintain the basis of their subsistence is inverted and subverted. This means that, instead of having the upper
moral ground, as poor people who have the right to a livelihood that can
support their families, the poor are burdened with the moral responsibility of
compensating for the excessive consumption of the more well-to-do in the
global South and particularly in the global North.

Our analysis draws on information collected as part of a larger study on
large-scale land investments, agriculture and food security.[4] Fieldwork took
place in the Mufindi district in June–August 2013 and included participatory
rural appraisal techniques (using focus group discussions (FGD) and key infor-
mant interviews),[5] structured and semi-structured interviews with both households and key informants (government organisations, NGOs and local
community structures),[6] and observations during field visits. Through consultations with the District Natural Resource Officer and the District Agricultural and
Livestock Development Officer the villages of Chogo and Mapanda in Mapanda
Ward were selected as field sites, based on their levels of large-scale land investments coupled with land use changes posing a potential threat to food crop production. As part of the study we also reviewed reports from district, regional
and national conservation land-use plans and policies, poverty reduction strategies and maps, as well as academic publications and reports on land acquisitions. Before turning to our analysis, we will first present the notion of moral
economy.

Moral economy

Environmental sustainability and the climate are, to use McDowell's words,
among the 'key moral issues of the early twenty-first century'.[7] These moral
issues are often couched in economic terms, involving an implied notion of
'moral economy' comprising 'some form of balance between ethical norms and
economic principles in order to achieve a degree of social justice'.[8] Climate
change mitigation measures, such as establishing forest plantations,[9] thus
ostensibly fulfil a global moral economy's requirements. They do so by bringing
economically sustainable investments to the global South while improving the
global environment, thereby garnishing support from donors and local
governments.

There is a long tradition of linking morality to economy because, as
Thompson, Scott, and others have shown,[10] the notion of a moral economy historically underpinned the moral rights evoked by the European peasants and
workers when they opposed perceived injustice, i.e. any action threatening the
social arrangements that enabled their subsistence. Today, Edelman comments,
the right to subsistence evoked by the economic underclass through the notion

of moral economy translates into the 'right to continue being agriculturalists'.[11] As Scott points out, this right is now being seriously challenged: 'The insecurities of smallholder agriculture have been amplified exponentially in this new world of IMF-structural adjustment loans; commodity dumping; intellectual property rights; new markets for credit, technology, and services; and giant agribusiness conglomerates.'[12] As we shall see in the Tanzanian case, an implied moral economy is used to justify land acquisition for forestation that will benefit private investors, international as well as national, at the expense of land traditionally used by local communities for subsistence. Viewed in this light, appeals to a global moral economy legitimating the enclosure of land used for subsistence agriculture can be seen to be an inversion of the classic understanding of the moral economy, which may also potentially have deleterious effects on the long-term subsistence needs of people.

After introducing land acquisition in our two study sites, we examine the historical and structural processes that have led to and enable present land acquisitions. We will then turn to our analysis of the narratives justifying land acquisitions for forest plantations and, finally, the local people's response to the land acquisitions in the light of these narratives.

Study sites

The study sites of the villages of Mapanda and Chogo are located in Mapanda Ward, Mufindi District in the Iringa Region, which is situated in the southern highlands of Tanzania (Figure 1).

According to the Iringa regional socioeconomic profile, more than 90% of Mufindi District's population live in rural areas and depend on agriculture as their major economic activity.[13] When research was carried out in 2013, there were 6765.9 km^2 of agricultural land, of which 1331.5 km^2 were cultivated. Seventy-seven per cent of the villagers interviewed responded that they owned land, mostly between one and 10 acres held in smaller plots scattered throughout the village. Their food crops were maize, sweet and white potatoes, beans and millet, while the cash crops were tea, coffee, pyrethrum and sunflower. Manufacturing industries comprised tea and pyrethrum processing and wood-related industries (including paper manufacturing and saw milling), while other industries included grain milling and brick making.[14]

Mapanda and Chogo are traditional villages that existed before 1978, when formal registration was carried out following the adoption of the Villages and Ujamaa [familyhood] Villages (Registration, Designation, and Administration) Act of 1975. The villagers are indigenous to the area, most being Wahehe (96.6%) with a small fraction from other ethnic groups, such as Wabena (2%) and Wakinga (0.7%).[15] A few immigrants (0.7%) have moved to Mufindi District from the neighbouring districts of Songea, Njombe and Kilolo to work in the tree planting and timber business, and as government officials employed in service sectors such as education, agriculture and governance.[16]

The district ranks first in the region in terms of forest cover thanks to increasing conservation and commercial forestry projects (Interview, District Natural Resource Officer, June 25, 2013). From 2001 until 2012 the number of trees planted annually rose from 3,662,816 to 31,174,500, the total number

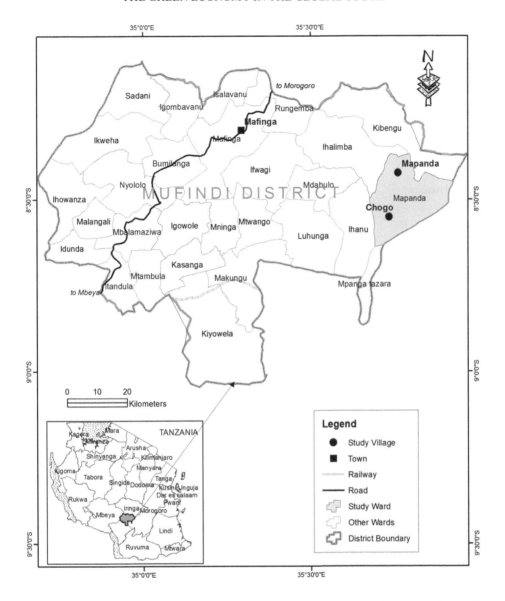

Figure 1.   Map of Mufindi District showing the study area. The insert shows the location of the Mufundi District within Tanzania.

reaching 191,371,183 in 2012, corresponding to an area of 119,606 ha.[17] Table 1 lists all large- and medium-scale investors whose activities use large tracts of land (50 ha and above) in the district.

Three large-scale investors are formally listed in the district profile as currently actively acquiring land in Chogo and Mapanda: Norwegian Green Resources Limited (GRL),[18] Fox Farms Highland Ltd and Duville Woodworks.[19] The latter two identify themselves as tourism and woodwork projects but fit squarely into the forest economy of other large-scale investors because of their direct involvement in large-scale forestation. GRL is the largest investor,

Table 1. Investments by type and size.

Name of investor	Category	Activity	Significance
Green Resource Ltd (GRL)*	Medium	Investment in tree planting, wood and carbon cess	Presence in Chogo and Mapanda village lands
Sao Hill Industries*	Medium	Investment in wood, poles and furniture	A subsidiary of GRL
MPM (Mufindi Paper Mills)	Large	Paper making	Involved in large-scale use of plantation forest products as raw materials
Sao Hill Plantation*	Large	Tree planting	Represents investors at the district level. Is also a central government investment
Unilever Tea (T) Ltd	Medium	Tea leaf processing	Represents investors at the district level
Pyrethrum Company of Tanzania (PCT) Ltd	Medium	Pyrethrum processing	Represents investors at the district level
Chai Bora Ltd	Medium	Investment in Tea blending and packaging	Represents investors at the district level
Mufindi Tea Company*	Medium	Tea leaf processing	Represents investors at the district level
Mufindi Wood Plantation & Timber Company	Medium	Investment in tree planting, wood and pole processing	Represents investors at the district level
Mufindi Poles and Timber Impregnation	Medium	Investment in wood and pole processing	Represents investors at the district level
Meena Wood Briquetting	Small	Wood briquetting	Represents investors at the district level
Tom Duville*	Small	Investment in poles, wood and furniture	Presence in Chogo and Mapanda village lands
Fox Farm Ltd*	Small	Investment in tree planting and tourism	Presence in Chogo village
Highland Forest Company (HFC)*	Small	Investment in tree planting, poles, wood and furniture	Currently dormant because of an existing border conflict with Mapanda village

*Included in the sample.
Source: MDC, *Mufindi District Socio-economic Profile*, 2010.

having 99-year leases on 6269 ha of land. This amounts to 27% of the total area of the two villages; most of this area, according to our observations, has already been planted with trees. However, there are also many individual investors (Tanzanians from other areas) who have acquired a total of 5% of the land in Chogo and planted trees on up to 1000 ha of land.

Neither the district nor village governments have a formal register of these individuals because land right transfers generally take place informally between the buyer and seller. It is only when land buyers require formal certification that they consider registering land through the village and district council. When obtaining formal certification, however, they will be identified as investors, obliging them to participate in village development projects. The informality of many land transactions means that large tracts of land under forest plantation are not officially recognised as investments. Our inquiries about these investments and obtainment of a list of those who have land in the village yielded unsatisfactory results, as village governments could provide only spec-ulative information except when the land buyer had provided information (even only informally) on the land transfer to the village office, as in the case of token compensation by investors (see discussion of Table 3).

Having briefly set the scene by describing the land acquisitions for forest plantations in our two study sites, we will in the following section further describe the different historical processes and structural factors enabling and leading to the large-scale land acquisitions we see in Africa today in general, and in Tanzania in particular.

Land acquisitions

'Investments', 'mitigation measures', 'productivity increase', 'grabbing', 'enclosure' – the phenomenon whereby (trans)national investors acquire large areas of land has many names and comes in different shapes and forms. Land acquisitions, using the neutral term, occurs for purposes ranging from efforts to ameliorate global food shortages to tourism to conservation and global environmental issues.[20]

The areas in which land acquisitions take place have often undergone several different kinds of land transfer that are still in effect, with varying potential benefits and challenges for the local communities. Some areas have been acquired by investors many years ago for the purpose of growing export crops, such as tea, while others have been acquired more recently and turned into sites for biofuels, trees or crops. During colonial rule land acquisitions were mainly driven by European colonialists; today, however, domestic elites and the African states are usually partners, intermediaries or beneficiaries in land acquisitions involving investors from the global North as well as South–South investments.

In relation to the role of the state, Evers et al explain that 'the state, in both his-torical and contemporary contexts, acts as a crucial actor in creating the fertile ground for foreign investments'.[21] The role of the state, however, is often inter-linked with the role of global actors, especially foreign donors. Evers et al explain:

> With the impetus of the World Bank, African governments assist foreign investors in engaging in a policy process of titling and commodifying land. Land reforms instigated by the state in compliance with imperatives of Foreign Direct Invest-ment (FDI) policies embraced by the World Bank have created an atmosphere favourable to large-scale land acquisitions.[22]

African states in this way facilitate large-scale land deals in the process of reasserting the state's 'power and identity'.[23] This is arguably necessary since high debt loads and the structural adjustment policies of the 1970s weakened many African states.[24] States thus actively use land acquisitions in order to reduce their weak position, and are arguably not merely 'passive victims'.[25] Land acquisitions can in some cases lead to personal economic benefits through rent seeking and corruption, or elite capture of land access and control.[26]

The Tanzanian state introduced land reforms in the 1990s, including new and modified laws promoting investment. Noe argues that this enabled de facto expropriation of customary land rights by re-defining most untitled (but traditionally occupied and used) lands, reinforcing that the state, by default, be their legal owner [...] Against this background, land tenure systems in Africa are fine-tuned to support the ongoing transformation of rural landscapes.[27]

New policies in Tanzania have further supported this transformation, in particular the 2009 policy of *Kilimo Kwanza* ("Agriculture First"), which, according to Smucker et al, makes 'land available for large-scale capital investment and production'. It does so partly through an amendment of the Village Land Act No. 5 of 1999, which makes it easier for investors to access village land.[28]

In Chogo and Mapanda, land may be obtained and transferred between villagers and newcomers through both customary and statutory laws. Traditionally land has been transferred from one person to another through local cultural practices, such as inheritance and informal borrowing of land from relatives. Today, however, land is increasingly acquired through purchase and rental agreements. Thus 17.4% of the respondents had purchased their land and 31.5% had rented land, while 57% had acquired land through inheritance and 16.8% had borrowed land. Statutory laws giving village councils the power to administer land in their jurisdiction (particularly the Village Land Act of 1999) allow village governments to (re)allocate to villagers or investors any land (not exceeding 50 ha) under their control, as needs arise.[29] Among villagers 8.1% responded that they had acquired land that had been allocated by the village government.

The village assembly is required to give its endorsement of land acquisitions above 50 ha for further processing in the district council. The democratically elected village government forms Tanzania's lowest governmental level. Within the district council, the village council is responsible for supervising and maintaining law and order, protecting public and private property and furthering the socioeconomic development of the village. The village assembly, comprising all adults above 18 years of age, is the only institution where villagers can raise issues and discuss them publicly.[30] Its agenda, however, must be approved by the village council before the meeting, thus limiting the villagers' ability to raise issues. The limited education of villagers and even councillors, as well as corruption and the power of the elite, are also likely to influence these discussions' direction and character.[31] This appeared to be the case in the study villages, where district officials (particularly land and legal officers), as well as prospective investors or their representatives, attended discussions of investment issues at village assemblies. The villagers often appeared to be intimidated by the dominating presence of officials favouring the investors. As noted by Smucker et al:

> The magnitude of the money and the size of these land deals makes it nearly impossible for local authorities to protect land rights against external threats [...] In addition, in some cases local authorities collude in alienating land from communities to enable private investment.[32]

While there are structural processes enabling land acquisitions, we argue here that narratives also play an important role in understanding the growth in land acquisitions. In the following we will move to our analysis and critique of the narratives justifying the land acquisitions, arguing that these narratives invert traditional suppositions concerning the moral economy.

Moral economies of green investments

The present scale of land acquisitions is unprecedented since the end of colonial rule. Indeed, the literature points to many similarities between colonial and contemporary practices, and the 'common wisdoms' justifying these practices. Based on an analysis of our empirical data and a literature review we present and problematise the key common wisdoms that constitute the foundation of the global moral economy narrative supporting the land acquisitions. We begin by focusing on the common wisdom that 'idle' land is 'unused', 'under-utilised' and a 'waste', and that investments in its improvement therefore saves it from neglect.

'Idle' land is waste

According to Deininger, Lead Economist in the World Bank's Development Research Group, there is a 'major scope for productivity increase on currently cultivated areas' and a need for 'identification of countries where demand for land expansion may concentrate'.[33] He adds that 'in virtually all African countries where demand for land acquisition has recently increased, the level of productivity achieved by existing (smallholder) cultivators is less than 25 percent of potential'.[34] Indeed, Evers et al state that Africa 'is often portrayed as the "lost continent": steeped in tradition, isolated from markets and trapped in the past – a place where vast swathes of land either lie unoccupied or are poorly utilized'.[35]

The notion of under-utilised and idle land, Gausset and Whyte point out, has legitimised land acquisitions since colonial times. Much of such land in colonial Africa, however, 'either formed part of a local land reserve in a system of shifting cultivation, or was used as commons for grazing, hunting, or collecting non-timber forest products'.[36] Today land availability is gauged largely in relation to (the lack of) agricultural productivity.[37] An assessment of agricultural productivity in the Mufindi District thus states that only 27% of arable land is under cultivation,[38] disregarding customary land use practices in the area. According to our informants, in Chogo and Mapanda large inherited tracts of land were traditionally kept for seasonal use, to regain fertility (eg fallowing in intervals of more than five years) and until the mid-1990s were used as a resource that could be given to a friend or needy relative. Furthermore, until recently the land was also considered family reserves that had cultural significance as ancestral graveyards, marking important ritual sites, and which could be siphoned off to future generations as part of their inheritance.

The idea that land was being ineffectively used was furthered by the socialist Villages and Ujamaa Villages Act of 1975,[39] which led to villagisation and compulsory *Ujamaa* villages. While Mapanda and Chogo existed before the Act, its promotion of nucleated settlement concentrations led to large areas of newly depopulated village lands being perceived and labelled as 'idle'.[40] Today President Jakaya Kikwete and other government officials promulgate this understanding, as they 'repeatedly state that 44 million of Tanzania's total land area of 94.5 million hectares are arable but only 23–24 per cent are being used'.[41]

Another way in which the notion of idle land is actively promoted by the state is through maps. The current interpretation of The 1999 Land Act and

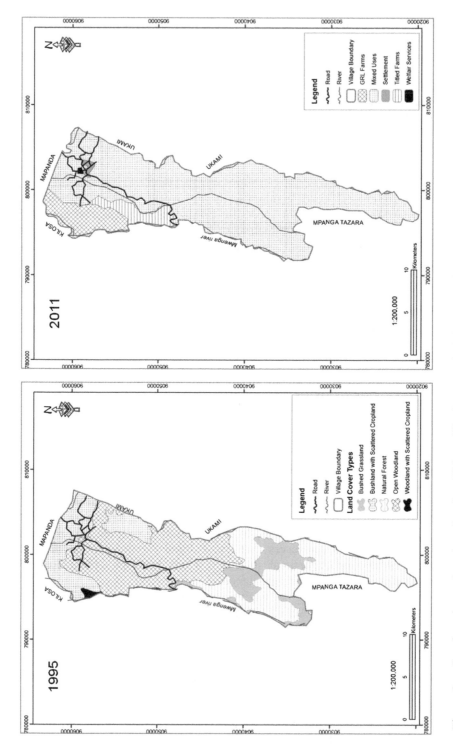

Figure 2. Land use/cover based on Landsat TM satellite imagery (1995) and the first formal land use plan in Chogo village (2011).

Village Land Acts supports the creation of private farms and plantations that are formally demarcated on village maps and endorsed by the president. The president has the power to transfer 'village land' to 'general land' before reallocation of such areas to private investors, leading to an irrevocable change of land tenure. Land thus changes from village council management to central government control. Land use plans, furthermore, are a legal requirement for formally registered investors, who normally will seek to acquire land legally. Indeed, facilitating the making of plans works to the advantage of investors, because this fast-tracks their activities and minimises conflicts with villagers and because layout maps are produced on the basis of these land-use plans.

Figure 2 presents land use for 1995 and 2011 in Chogo village. Whereas the 1995 map was generated from Landsat TM satellite images, the 2011 map was retrieved from the village land-use planning maps and is the first formal land-use plan. This plan, which enables land transfers, was proposed and paid for by GRL. According to the District Natural Resource Officer, the primary involvement of the district and village councils was to mobilise support in the villages and fulfil the regulatory requirements of participatory land use planning (Interview, July 2, 2013). As can be seen in Figure 2, there are no forest plantations in 1995, and natural vegetation covers most of the village land; scattered croplands and settlements are concentrated in the northern parts, while southern parts are used seasonally. In the 2011 formal land-use plan all natural vegetation has been compiled into a vague 'mixed use' category (84% of the land) which includes village tree farms and croplands. The map furthermore includes categories representing large tree plantations, thereby recognising them formally, such as 'GRL' and private 'titled farms' comprising 15% of the village.

Village land-use plans help ensure sustainable land-use practices, food production and security, and guide land ownership transfers to investors. Village land-use plans are therefore currently considered to be part of the positive results of foreign investments, as investors often help finance them, as in the case of GRL. These plans are nevertheless often profit-driven, serving to facilitate land-right transfers. Among the seven villages in Mapanda Ward, only the two study villages where there are large-scale investors like GRL have land use plans. Powerful participatory planning narratives, overshadowing possible criticisms from within and outside the rural community, dominate government and investor documentation of the planning process. Accordingly, investor-driven village land-use plans raise questions concerning fairness in setting priorities between local development and investor interests. Knowledge of how land tenure is changed in the process of transferring land from villagers to investors with large-scale foreign capital is thus an important, if underplayed, issue in the assessment of the impacts of these investments on food security and other development aspects. Our study suggests that village land use plans are not ends in themselves but part of a process bringing irrevocable changes to village land tenure and ownership, further highlighting the state's role in facilitating land acquisitions.

Global investments lead to local benefits

The second 'common wisdom' – that outside investment generates local benefits – also has deep historical roots. Under colonial rule land acquisitions by outsiders were legitimised as bringing investment, technologies, employment and 'higher living standards, as well as food, to a growing world population'.[42] Today, Deininger similarly argues that:

> investments can provide benefits through four channels: (i) social infrastructure, often supported by community development funds using land compensation; (ii) generating employment and jobs; (iii) providing access to markets and technology for local producers; and (iv) higher local or national tax revenue.[43]

This influential narrative emphasising the local benefits of outside investments, has been questioned, however.

The employment opportunities supposedly generated by land acquisition have been severely criticised in the general literature. Focusing on Southeast Asia, Li explains:

> plantations have routinely been bad news for the 'locals': their land is needed, but their labour is not. By selecting areas with low population density, managers can argue that labour is in short supply so they must import it. Significantly, the people whose land is taken over by the plantation are seldom employed there, a practice legitimated by the 'myth of the lazy native' [...] but more accurately a reflection of the difficulty of extracting consistent, cheap labour from people who still have access to patches of land in the vicinity, hence other options.[44]

As a result, 'the optimal configuration, from a profit-making perspective, is one in which labour is superabundant, hence cheap and easily disciplined'.[45]

As Li explains, with respect to the profitability of Sumatra's plantation belt, initiated in the 1920s, this was generated by limiting the land of the local population, and then waiting for the population increase to be so large that the land available was

> insufficient to sustain the increased population stuffed in the nooks and crannies between the plantations, hence they were obliged to work for the plantations but on adverse terms, as 'temporary' contract workers, further disciplined by means of piece rates, netting pay far below the official minimum wage.[46]

Despite such historical precedent, 'states competing against each other to attract investors must be prepared to offer land at a competitive price, which often means free of charge, a move they justify with reference to other purported development benefits, especially jobs'.[47] While investors may provide jobs for the local population, they are not contractually committed to do so, and not committed to provide any particular conditions and wages for doing so. Furthermore, were the investors to leave, the local population would lose its source of income.

In Tanzania the notion that global investments generate local benefits is used by local governments and investors alike to justify the financial support for international investors from international donors and NGOs. GRL, for example, has thus received funding from the Norwegian Agency for Development Cooperation (NORAD).[48] This is also exemplified in Tanzania by the *Kilimo Kwanza* policy, which seeks to generate foreign direct investment (FDI) to promote large-scale agribusiness and also, in parallel, to transform small-scale farmers into commercial small farmers on the model of the Green Revolution. Indeed the *Kilimo Kwanza* initiative has won substantial financial support from the alliance for a Green Revolution in Africa (AGRA).[49]

Smucker et al cite the newspaper, *The Citizen*, as criticising this policy for turning 'small-scale farmers into labourers in their own country'.[50] Furthermore 'Civil society and academic critics have suggested that the stated concern for poor farmers belies the emphasis of *Kilimo Kwanza* on ensuring land access for large scale, export-oriented production'.[51]

Despite such concerns, the benefit of land-based investments to the local population is highlighted by investors such as GRL. Thus, on the welcome page of the GRL website, they highlight four key benefits: forests, wood products, carbon credits and community development. Concerning the latter area they state: 'Green Resources facilitates socio-economic development and poverty alleviation in rural areas through provision of employment, infrastructure development, schools, health and other community development' (http://www.greenresources.no/, accessed July 21, 2014). Nevertheless, as we will discuss further below, the benefice of these ostensible good works has been questioned.

At the time of our fieldwork few villagers were actively working for the plantation, or participating in activities related to forestry. Traditionally every family in Chogo and Mapanda is expected to be able to produce its own food and 100% of respondents to our questionnaire said that crop cultivation was a major economic activity. Families do engage in other emerging economic activities, such as timber-related businesses, casual labour, professional employment, etc, especially to supplement subsistence production and the income from cash crops. But, as can be seen from Table 2, few are involved in activities like timber trading (2%) and lumbering (2.7%) that depend directly on forests,[52] and only 6.7% were casual labourers in private plantations.[53]

Because of the strong focus on the local benefits of foreign investment, outside investors can often make do with giving what amounts to token compensation for land acquired, eg by putting up schools, rather than paying for the land.[54] Support of this kind to the sectors of education, health and employment (see Table 3) exemplifies this type of compensation in our study area, where, thanks to its visible character, it can initially generate positive local perceptions

Table 2. Villagers' main economic activities.

Activity	Counts (n=149)	Percentage of respondents
Crop cultivation	149	100.0
Livestock keeping	57	38.3
Professional employment	5	3.4
Making/selling charcoal/firewood	2	1.3
Making and selling local brew	3	2.0
Casual plantation labourer	10	6.7
Lumbering	4	2.7
Timber business	3	2.0
Bee keeping	24	16.1

Table 3. Investor land acquisitions in Chogo and Mapanda.

Investor	Year acquired	Village	Land size (ha)	Purpose	Land costs /compensation	Current use
GRL	31/08/1997	Chogo	8000	Trees	Roofing school teacher's house, providing transport to villagers when needed	Forest plantation
Mrs M.S. Fox	20/10/1999	Chogo	1500	Coffee and dairy farm	Employment, roads, water wells, school renovations, training villagers on coffee and dairy farm	Forest plantation
Major Gen James Luhanga	23/10/2000	Chogo	400	No specific land use (mostly unused)	Participation in village development	Sold/transferred to GRL
Pascal Mhongole	23/10/2000	Chogo	1000	Coffee farm	Coffee processing machine	Sold/transferred to GRL
Augustine Pascal	12/11/2000	Chogo	40	Trees	No compensation	Trees
GRL	12/08/2006	Mapanda	2	Tree nursery	No compensation	Tree nursery
HFC	24/04/2008	Mapanda	20	Trees	Two classrooms in the school	Air strip and tree nursery
Mrs M.S. Fox	03/01/2010	Chogo	750	-	Books and pens for primary school	Trees

Source: Minutes of village meetings, Chogo and Mapanda.

of the socioeconomic impact of large investment companies. This is despite the fact that the compensation comes in the form of charity rather than a contractual binding agreement between equal partners. Furthermore, since there is no binding contract, some investors do not live up to their promises. Refseth thus found that, apparently because of the low economic performance of the plantation, GRL was 'lagging behind in fulfilling their promises made during the acquisition' and that 'no documents regarding the promises' existed.[55] Another problem she noted was that GRL promises that 10% of the sale of carbon credits would go to community projects, thereby using "new" money to fulfil "old promises."[56]

Interestingly, when asked who benefited most from the village's natural resources, 46.8% of villagers responded that it was large-scale investors; 22.2% responded that it was those with financial capacity and only 9.3% responded that it was both villagers and investors. When asked about their general views on the presence of the investors, about 52% of the respondents were of the opinion that investors should continue but with closer monitoring regarding the promises they had made.

Trees are 'green'

The third 'common wisdom' behind the global moral economy narrative, that trees are 'green', is reflected in the international framing of forest plantations. Thus the plantations are not only seen to bring socioeconomic development to local people, but also viewed as beneficial to the global environment. These arguments echo similar arguments made at the time of 18th century European land acquisition through enclosure, when common lands were termed 'waste' and afforested in the name of both economic and environmental improvement, despite the moral economic protests of the commoners. These arguments are still current, for example in the use made by modern land appropriators of evidence supposedly supporting the 'tragedy of the commons' thesis. This evidence, which was built upon 18th century arguments, is challenged by those who argue for the viability of commons.[57]

Forest plantations thus address key moral issues of our time which have a long history, namely socioeconomic and environmental sustainability.[58] For example, GRL has been given Forest Stewardship Council (FSC) certification and Voluntary Carbon Standard (VCS) certification.[59] Furthermore, GRL encourages villagers to plant trees on their own land. Villagers, according to our findings, have embraced tree growing as a viable economic activity, partly because of the carbon trade. In fact, in 2008 the vice president's office identified Mufindi district as one of the National Carbon Credit pilot areas based on its rate of forest cover. Chogo thus received 10 million Tanzanian shillings (Tshs) and Mapanda 30 million Tshs in 2013 from GRL from the carbon trade.[60] Interestingly, as Stave explains, 'For Norway, the establishment of such a market will be of great interest, since it is considered economically unrealistic for the country not to exceed its Kyoto assignment'.[61] Some tree species are, however, very hard on the land. The GRL plantations primarily consist of different subspecies of pinus and eucalyptus that can lead to local environmental problems such as nutrient depletion and water deficiency.[62] Run-off reduction is particularly a problem when grasslands and scrublands are being afforested.[63] Indeed, 21.4% of the villagers mentioned water scarcity as their main environmental concern, while 44.9% described a trend of decreasing water resources over the past 10 years. Furthermore, large-scale monoculture forest plantations reduce local biodiversity.[64]

Villagers can earn income from tree production but this involves a reduction not only in available farm land, as trees occupy land for many years until they can be harvested, but also in time available for farming. Discussing a Mexican case, Osborne notes: 'Carbon enclosures operate by limiting traditional farming practices on previously worked land and are policed through environmental monitoring by local NGOs and international institutions'.[65] About 50% of villagers in our case areas anticipated having to adopt forestry as their future primary land use. Only 10.7%, however, listed tree planting as their preferred future livelihood option, and 18.1% listed forest increase as a future environmental concern. Apart from productivity loss as a result of loss of labour needed for the family plots, new emerging forestry-related threats to the subsistence of the local population include the creation of suitable habitats for problem animals such as monkeys and warthogs, which cause crop damage. Thus 25.4% of villagers responded that an increase in wild animals was a future environmental concern. When asked how environmental challenges could be addressed, 23.5% responded that the number of investors should be reduced and 20.2% responded that land should be set aside for crop production.

The current rate of forestation of family plots, whereby food security could be jeopardised in different ways in the near future, illustrates the dilemma inherent in attempts to ensure both global and local food security through a 'moral' economy. What happens to local food security when land is being acquired in order to compensate for climatic threats to global food insecurity? Borras et al comment that policies enabling such land acquisitions: 'foreground "security" for some, while leaving others without shelter, food, or the means of (re)production'.[66] Smucker et al add that:

> concerns have been further highlighted in the advent of new climate change mitigation projects, such as United Nations REDD+. Implementation of REDD+ could mean that the large proportion of Tanzania's rural population that lacks land certificates could be excluded from the benefits of climate mitigation projects that conserve forest resources over which they hold only customary rights.[67]

The moral economy of the local poor?

According to district and village government statistics, the Mufindi area maintains an 'average production' of food crops, suggesting that there have not been major changes in this respect during the past 10 years.[68] While generally confirming that few changes had occurred in the actual amount of cereals produced by households, 55.7% of villagers reported food gaps during part of the year, and 4% throughout the year. Furthermore, about 50% of respondents felt their food security status had worsened, largely because households had to sell cereals after harvests to pay for such things as school fees, health services, better housing and transport. Hence food security was threatened by the lack of alternative cash sources. This has reinforced the view among farmers that invest-ment projects that promise to provide such sources of income, including employment opportunities, as well as reducing dependence on subsistence farming, are key drivers of rural development.

As a result of such promises, it appears as if the local farmers have been pacified, and resistance quelled, the narrative of a global moral economy effec-tively overwriting the moral economy of the poor. At the moment the repercus-sions of the land acquisitions and the change in food security may not be entirely clear to the villagers as a result of incomplete information. This may explain the present lack of resistance. If, however, climate change does indeed create new challenges in terms of reconfiguring the possibilities for food produc-tion, it would seem to be of the utmost importance for the local populations to have access to their land reserves, or so-called idle lands, in order to optimise their climate change adaptation options. As this statement by one of the villagers in Mapanda indicates, this situation may lead to an unacceptable burden:

> Trees are not bad; they help with reducing impacts of climate change. However, exchanging trees for food is not going to work. People are not oriented towards money economy – so they may have money but still be food insecure unless extensive training is done on cultural changes. Cultural change is not a one-time thing, it can take a generation. It seems that during these changes, villagers will completely depend on investors to feed themselves, that is, they only have labour to exchange for food. What a bad turn of things! (Interview, Maurius Kisinga, June 27, 2013)

Traditional community livelihoods are not oriented towards a money economy where food is budgeted for (as opposed to being produced on family farms). The observable changes from a traditional to an exchange economy in the study villages represent primarily conditions that generate the fundamental supply of money-capital to facilitate the functioning of the market. As indicated by the quote above, villagers worry they may have a hard time adapting to this. Never-theless, 69.1% of respondents considered professional employment as among their preferred future livelihood options and 86% of villagers listed buying as a means of accessing resources. There is thus already a very observable change from an exchange economy which, at least in the short run, as discussed above, is leading to villagers having to sell cereals, and thereby reducing their food security. This may also be a factor in why villagers are planting trees on their lands at the expense of crops, even though few wish tree planting to be their future livelihood option. The ambiguity the villagers feel towards this change is perhaps also reflected in their response to the trends they have observed in resources over the past 10 years: 91.9% responded that there had been a decrease in agricultural land resources and 59.1% had observed a loss of cultural resources, with 38.5% listing land scarcity as a major socioeconomic constraint.

Conclusion

In the Mufindi district customary practices of land-right transfers between generations have met global forces that support commercial forestry, supplanting customary uses of land for crop production. As a result of the growing trend toward forest plantations owned by private companies, the central government and individuals, land coverage under plantations has increased steadily over the past 10 years, with subsequent impacts on land use. Notably the past 10 years have seen over 30% of the land in the two study villages being converted into private properties to which villagers will have no access, nor control over the land, for 99 years. These investments will have multifaceted effects on subsistence strategies, particularly when considering the long-term transformations of livelihoods, land rights, access and ownership. By losing access to land for 99 years the local populations are moving towards a cash-based economy, where employment in the plantations or even emigration to distant destinations for wage employment become crucial to subsistence, and life as an agriculturalist may no longer be viable.

Governments, donors and investors often promote land acquisitions for forest plantations as climate change mitigation via carbon sequestration. Investors' forestry activities thus become part of an imagined global moral economy. Using examples from a study focusing on two Tanzanian villages, we have critically examined the imagined foundation of the global moral economy narrative, which presents trees as axiomatically 'green', 'idle' land as waste and economic investments as benefiting the relevant communities. We have found that this moral justification subverts contractual commitments towards compensating local livelihoods for land lost, thereby enabling investors to trade land for charity. The traditional supposition of a moral economy as a system of norms and obligations invoked by the economic underclass to maintain their right to subsistence, in this case their right to be agriculturalists, is thereby reversed. Instead, investments are promoted that do little to minimise the excesses of the elites in the global North and South, while at the same time potentially undermining the subsistence strategies of the economic underclass of people whose land is taken in the name of a common moral responsibility to ameliorate these 'global' excesses.

Acknowledgements

We would like to thank Uma Kothari for insightful comments on an earlier draft and acknowledge the important contribution to this paper made by participants at the conference 'Green Economy in the South: Negotiating Environmental Governance, Prosperity and Development' held at University of Dodoma, July 2014. In particular we thank our co-presenters Alex Dorgan and Flora Hajdu, as well as the audience, for very helpful comments and questions, and Tor A Benjaminsen for his suggestions, which substantially shaped the revisions to this paper. Additionally, we thank Mufindi District officials, representatives of investment companies, village leaders as well as villagers for their invaluable cooperation. Finally, we are grateful for the financial support received from Danida BSU-GEP, without which this study would not have been possible.

Funding

This work was supported by the Danida Fellowship Centre BSU-GEP, [project no. 32663].

Notes

1. Locher and Sulle, *Foreign Land Deals*, 36.
2. Ibid.
3. For example, Leach and Mearns, *The Lie of the Land*.
4. Kangalawe et al., *Entailments of Large-scale Land Investments*.
5. Individual interviews were initially conducted with different government department heads, who assisted in organising FGDs comprising other relevant officials. In total three FGDs were conducted in district council departments; in the District Natural Resource Office, District Land Office, District Planning Office and District Agricultural and Livestock Development Office. In the villages respondents for six in-depth interviews were identified through purposive sampling. One village council FGD was conducted in each village and two additional FGDs in Chogo village with villagers working in the plantations.
6. A structured questionnaire was administered to a sample of randomly selected households: 102 households in Mapanda and 47 in Chogo, representing about 10% of the village households. There are some gaps in the statistics as the available 2012 census report does not provide information about village population but only about ward population. Drawing a 10% sample was therefore done based on the village registry of households and using sub-village leaders who knew the exact numbers of households in their areas. That is, the 10% were drawn from the level of the sub-village.
7. McDowell, "Moral Economies," 189.
8. Ibid., 187.
9. The focus of this article is on forest plantations; however, conservation and REDD+ are also pertinent examples.
10. Thompson, "The Moral Economy"; and Scott, *The Moral Economy of the Peasant*.
11. Edelman, "Bringing the Moral Economy back in," 332.
12. Scott, "Afterword to 'Moral Economies'," 397.
13. URT, *Iringa Region Socio-economic Profile*.
14. Ibid.
15. MDC, *Mufindi District Socio-economic Profile*, 2012.
16. MDC, *Mufindi District Socio-economic Profile*, 2011.
17. MDC, *Mufindi District Socio-economic Profile*, 2013.
18. GRL is a subsidiary of Green Resources AS.

19. Both run by expats.
20. Borras et al., "Towards a Better Understanding"; and Gausset and Whyte, "Climate Change and Land Grab."
21. Evers et al., "Introduction," 5.
22. Ibid. See also Noe, *Contesting Village Land*.
23. Evers et al., "Introduction," 17.
24. Evers et al., "Introduction"; Wolford et al., "Governing Global Land Deals"; and Noe, *Contesting Village Land*.
25. Wolford et al., "Governing Global Land Deals," 192.
26. Ibid; and Evers et al., "Introduction."
27. Noe, *Contesting Village Land*, 4; See also Shivji, "Not yet Democracy"; and Alden Wily, "'The Law is to Blame'."
28. Smucker et al., "Differentiated Livelihoods," 44.
29. URT, "The Village Land Act."
30. Grawert, *Departures from Post-colonial Authoritarianism*.
31. Ibid; and Shivji, *Village Governance*.
32. Smucker et al., "Differentiated Livelihoods," 42.
33. Deininger, "Challenges Posed," 217.
34. Ibid., 218.
35. Evers et al., "Introduction," 13.
36. Gausset and Whyte, "Climate Change and Land Grab," 218.
37. Cotula et al., *Land Grab or Development Opportunity?*; Shivji, *Accumulation in an African Periphery*; Alden Wily, "Looking Back;" and Noe, *Contesting Village Land*.
38. MDC, *Mufindi District Socio-economic Profile*, 2010.
39. URT, *The Villages and Ujamaa Villages*.
40. While land ownership did not change as a result of villagisation, and families retained most of their use rights to land, villagers have difficulty getting these rights recognised, partly because of an inability to access appropriate legal institutions. Cotula, "The International Political Economy."
41. Askew et al., "Of Land and Legitimacy," 120.
42. Gausset and Whyte, "Climate Change and Land Grab," 220.
43. Deininger, "Challenges Posed," 226.
44. Li, "Centering Labor," 286.
45. Ibid.
46. Ibid., 287. This critique is very similar to Marx's notion of 'primitive accumulation' – 'the historical process of divorcing the producer from the means of production' – referring to the enclosures of the commons in England as the classic example. Marx, cited in Benjaminsen and Bryceson, "Conservation," 336. Building on the idea of primitive accumulation, Harvey has coined the term 'accumulation by dispossession' to describe the situation today. Harvey, cited in Benjaminsen and Bryceson, "Conservation," 336.
47. Li, "Centering Labor," 283.
48. Green Resources has benefited from NORAD funding in several ways. For example, the company took over Sao Hill Sawmill in 2003, which at the time of the take-over was privatised, but had originally been financed with NORAD funding. Refseth, "Norwegian Carbon Plantations," 51.
49. Smucker et al., "Differentiated Livelihoods," 44.
50. Ibid.
51. Ibid.
52. People engaging in timber-related business, however, commute between villages and town centres, and may therefore not have been present during the time of this study. Villages are nevertheless largely suppliers of logs, and economic activities related to saw milling are yet to substantially have an impact on the village economy.
53. The small number involved in forestry-related activities could partially be a result of a lack of timber processing centres in the villages, which are difficult to establish as there is no electricity there.
54. Locher and Sulle, *Foreign Land Deals*.
55. Refseth, "Norwegian Carbon Plantations," 72–73. See also Benjaminsen et al., "Conservation."
56. Refseth, "Norwegian Carbon Plantations," 73–74.
57. Gausset and Whyte, "Climate Change and Land Grab."
58. McDowell, "Moral Economies."
59. Refseth, "Norwegian Carbon Plantations," 53.
60. Equivalent to US$5993 and $17,980, respectively.
61. Stave, "Carbon Upsets," 56.
62. Ibid; and Zinn et al., "Soil Organic Carbon."
63. Farley et al., "Effects of Afforestation."
64. Purdon, "Land Acquisitions in Tanzania."
65. Osborne, "Carbon Forestry," 880.
66. Borras et al., "Towards a Better Understanding," 214.

67. Smucker et al., "Differentiated Livelihoods," 42.
68. MDC, *Mufindi District Socio-economic Profile*, 2012.

Bibliography

Alden Wily, L. "'The Law is to Blame': The Vulnerable Status of Common Property Rights in Sub-Saharan Africa." *Development and Change* 42, no. 3 (2011): 733–757.

Alden Wily, L. "Looking Back to see Forward: The Legal Niceties of Land Theft in Land Rushes." *Journal of Peasant Studies* 39 (2012): 751–775.

Askew, K., F. Maganga, and R. Odgaard. "Of Land and Legitimacy: A Tale of Two Lawsuits." *Journal of the International African Institute* 83, no. 1 (2013): 120–141.

Benjaminsen, T., I. Bryceson, F. Maganga, and T. Refseth. "Conservation and Land Grabbing in Tanzania." Paper presented at the International Conference on Global Land Grabbing, Institute of Development Studies, University of Sussex, April 6–8, 2011.

Benjaminsen, T. A., and I. Bryceson. "Conservation, Green/Blue Grabbing and Accumulation by Dispossession in Tanzania." *Journal of Peasant Studies* 39, no. 2 (2012): 335–355.

Borras, S. M., R. Hall, I. Scoones, B. White, and W. Wolford. "Towards a Better Understanding of Global Land Grabbing: An Editorial Introduction." *Journal of Peasant Studies* 38, no. 2 (2011): 209–216.

Cotula, L. "The International Political Economy of the Global Land Rush: A Critical Appraisal of Trends, Scale, Geography and Drivers." *Journal of Peasant Studies* 39, nos. 3–4 (2012): 649–680.

Cotula, L., S. Vermeulen, R. Leonard, and J. Keeley. *Land Grab or Development Opportunity? Agricultural Investment and International Land Deals in Africa.* London: IIED/FAO/IFAD, 2009.

Deininger, K. "Challenges posed by the New Wave of Farmland Investment." *Journal of Peasant Studies* 38, no. 2 (2011): 217–247.

Edelman, M. "Bringing the Moral Economy back in...to the Study of 21st-century Transnational Peasant Movements." *American Anthropologist* 107, no. 3 (2005): 331–345.

Evers, S., C.Seagle, and F.Krijtenburg. "Introduction: Contested Landscapes – Analysing the Role of the State, Land Reforms and Privatization in Foreign Land Deals in Africa." In *Africa for Sale? Positioning the State, Land and Society in Foreign Large-scale Land Acquisitions in Africa*, edited by S. Evers, C. Seagle and F. Krijtenburg, 1–36. Leiden: Brill, 2013.

Farley, A. K., E. G. Jobbágy, and R. B. Jackson. "Effects of Afforestation on Water Yield: A Global Synthesis with Implications for Policy." *Global Change Biology* 11 (2005): 1565–1576.

Gausset, Q., and M. Whyte. "Climate Change and Land Grab in Africa: Resilience for Whom?" In *Climate Change and Human Mobility: Global Challenges to the Social Sciences*, edited by K. Hastrup and K. F. Olwig, 214–234. Cambridge: Cambridge University Press, 2012.

Grawert, E. *Departures from Post-colonial Authoritarianism: Analysis of System Change with a Focus on Tanzania.* Frankfurt: Peter Lang, 2009.

Kangalawe, R., C. Noe, E. Luoga, and M. F. Olwig. *Entailments of Large-scale Land Investments on Agriculture and Food Security in Mufindi East, Tanzania.* Report for Danida. Copenhagen: Ministry of Foreign Affairs of Denmark, 2013.

Leach, M., and R. Mearns (eds.). *The Lie of the Land: Challenging Received Wisdom on the African Environment.* Oxford: James Currey, 1996.

Li, T. M. "Centering Labor in the Land Grab Debate." *Journal of Peasant Studies* 38, no. 2 (2011): 281–298.

Locher, M., and E. Sulle. *Foreign Land Deals in Tanzania: An Update and a Critical View on the Challenges of Data (Re)production.* LDPI Working Paper 31. The Hague: The Land Deal Politics Initiative, 2013.

McDowell, L. "Moral Economies." In *International Encyclopedia of Human Geography*, edited by Rob Kitchin and Nigel Thrift, 185–190. Oxford: Elsevier, 2009.

Mufindi District Council (MDC). *Mufindi District Socio-economic Profile.* Mafinga: MDC, 2010.

MDC. *Mufindi District Socio-economic Profile.* Mafinga: MDC, 2011.

MDC. *Mufindi District Socio-economic Profile.* Mafinga: MDC, 2012.

MDC. *Mufindi District Socio-economic Profile.* Mafinga: MDC, 2013.

Noe, C. *Contesting Village Land: Uranium and Sports Hunting in Mbarang'andu Wildlife Management Area, Tanzania.* LDPI Working Paper 15. The Hague: The Land Deal Politics Initiative, 2013.

Osborne, T. M. "Carbon Forestry and Agrarian Change: Access and Land Control in a Mexican Rainforest." *Journal of Peasant Studies* 38, no. 4 (2011): 859–883.

Purdon, M. "Land Acquisitions in Tanzania: Strong Sustainability, Weak Sustainability and the Importance of Comparative Methods." *Journal of Agricultural and Environmental Ethics* 26 (2013): 1127–1156.

Refseth, T. H. D. "Norwegian Carbon Plantations in Tanzania: Towards Sustainable Development?" Masters' thesis, Norwegian University of Life Sciences, 2010.

Scott, J. C. *The Moral Economy of the Peasant: Rebellion and Subsistence in Southeast Asia.* New Haven, CT: Yale University Press, 1976.

Scott, J. C. "Afterword to 'Moral Economies, State Spaces, and Categorical Violence'." *American Anthropologist* 107, no. 3 (2005): 395–402.

Shivji, I. *Village Governance and Common Pool Resources in Tanzania.* Common Pool Resource Policy Paper 3. Cambridge: Department of Geography, Cambridge University, 2002.

Shivji, I. *Not yet Democracy: Reforming Land Tenure in Tanzania.* London, Dar es Salaam: International Institute of Environment and Development, HAKIARDHI and the Faculty of Law, University of Dar es Salaam, 1998.

Shivji, I. *Accumulation in an African Periphery: A Theoretical Framework.* Dar es Salaam: Mkuki na Nyota Publishers, 2009.

Smucker, T. A., B. Wisner, A. Mascarenhas, P. Munishi, E. E. Wangui, G. Sinha, D. Weiner, C. Bwenge, and E. Lovell. "Differentiated Livelihoods, Local Institutions, and the Adaptation Imperative: Assessing Climate Change Adaptation Policy in Tanzania." *Geoforum* 59 (2015): 39–50.

Stave, J. "Carbon Upsets: Norwegian 'Carbon Plantations' in Tanzania." In *Tree Trouble,* edited by Friends of the Earth International in cooperation with the World Rainforest Movement and FERN, 52–57. Oslo: NorWatch/Future in Our Hands, 2000.

Thompson, E. P. "The Moral Economy of the English Crowd in the Eighteenth Century." *Past and Present* 50 (1971): 76–136.

United Republic of Tanzania (URT). *Iringa Region Socio-economic Profile.* National Bureau of Statistics (NBS) and Iringa Regional Commissioner's Office. Dar es Salaam: The President's Office, Planning and Privatisation, 2005.

URT. *The Villages and Ujamaa Villages (Registration, Designation and Administration) Act, no. 21 of 1975.* Dar es Salaam: Government Printers, 1975.

URT. *The Village Land Act (No. 5).* Dar es Salaam: Government Printers, 1999.

Wolford, W., S. M. Borras Jr, R. Hall, I. Scoones, and B. White. "Governing Global Land Deals: The Role of the State in the Rush for Land." *Development and Change* 44, no. 2 (2013): 189–210.

Zinn, Y. L., D. V. Resck, and J. E. da Silva. "Soil Organic Carbon as Affected by Afforestation with Eucalyptus and Pinus in the Cerrado Region of Brazil." *Forest Ecology and Management* 166, no. 1 (2002): 285–294.

Performativity in the Green Economy: how far does climate finance create a fictive economy?

[a]*University of KwaZulu-Natal, Durban, South Africa;* [b]*School of Environment, Education and Development, University of Manchester, Manchester, UK*

This paper asks how far performativity in the Green Economy generates material or virtual assets. It examines the relationship between assets and their financial derivatives, asking how far the value of 'carbon' or 'green' can be directly attributed to its social and narrative construction. The paper draws on two case studies – one of the Clean Development Mechanism (CDM) in South Africa, the other of the global private green bonds market – to show that both public and private climate finance can generate virtual economic activity co-produced by processes of social valuation and accumulation proper. How reliant is the Green Economy on actual economic activity?

Introduction

This paper will apply writing from critical political economy and economic anthropology to understand the relationships between nature-based fixed assets, the emerging Green Economy and the wider global political economy of which it is a part. Since the empirical material for this is potentially vast but also subject to problems of accessibility, I will illustrate my argument using a single case study of the economic activities in South Africa financed from the Clean Development Mechanism (CDM) of the UN Framework Convention on Combatting Climate change (UNFCCC) between 2007 and 2012, this last date being when South Africa 'graduated' from the mechanism. While this is a relatively small case study in the wider political economy of climate finance, it does illustrate the great distance between the fixed assets, their derivative income streams and the potential centrality of performativity and spectacle to the Green

Economy assemblage. The theoretical question which arises is how far does performativity provide for a distal, or non-substantial, relationship between an asset and its traded derivative income? Can we predict when the distance is too great or indeed when fixed assets at scale are so degraded, or simply non-existent, as to cause a wider rupture or crash? At the very least I will attempt to theorise how values in the climate-financed economy as a whole are made or conditioned by the economic activities from which they derive.

In order to answer these questions it is necessary to contextualise the Green Economy within wider global economic relations, since it is being designed, built and/or performed within ever-greater financialised relationships between people, things and other species.[1] The need to conceptualise the Green Economy at global scale is imperative, given the relative failure thus far of climate finance to grow in relation to the required needs of climate change mitigation and adaptation in terms of the environment,[2] and changes required to human-built environments for a sustainable future.[3] Castree and Christophers recently discussed this problem and the viability of finance capital to perform a massive 'capital switch' in favour of a climate mitigating, climate adapting, new socioeconomic reconfiguration which rewrites humans' relationship with ecology,[4] reminding us of the growing evidence of the urgent necessity to do so.[5] The Green Economy is widely viewed and mobilised as the vehicle or in Callon's terms 'socio-technical arrangement' that would seem best designed to adopt this role.

The paper will ask how far valuation systems in climate finance, and their institutional conduits of, first, the CDM and, second, its successor the green bond market (as components of the Green Economy), are facilitating this type of capital investment at scale.[6] This implies a second work of theory, to understand how far the framing of 'green' in both the private sector and in publically funded climate finance mobilises substantive changes in the economy towards a greener and more socially just capitalism. This work of both theoretically exploring and empirically assessing 'green' investments, or 'drilling down' on climate-financed activities to date, is important and urgent. This is so in part because the overall Green Economy is framed in spectacle;[7] in part because it is empirically operationalised in secrecy jurisdictions and opaque financial vehicles;[8] and because studies of it have rarely disaggregated these practices into verifiable substantive activities in actual societies and ecosystems.[9]

The next section will further develop the conceptual basis for this exploration. The third section will empirically examine what can be seen to constitute 'green' economy assets, and review the South African case study. In the fourth section the relationship between the carbon economy, increasing green bond activity and the 'Green Economy' will be explored. In the subsequent two sections a theorisation of the political economy of valuation and of the real and the performed will be outlined, respectively. The argument is that a traditional Marxism that insists on a relationship between a fixed asset and ever-expanding derivative liquidity, which is seen, at least at some point, to lead to a crash, tells us little about the temporal, geographical or political moments that make and break that relationship. This is a point generic to critical work on capitalism in general, but theorising the Green Economy highlights this theoretical deficiency: it suggests that the relationship between a fixed asset and a derivative income stream from it can be stretched to the point of non-association. A final section concludes.

The article argues that a 'fourth moment' of accumulation is emerging (and the Green Economy is a good keyhole through which to view it), in addition to Harvey's classic spatial, temporal and displacement/expropriation moments.[10] Using the Green Economy as exemplar, I argue that this fourth moment is the performativity of value variously theorised by Callon, Miller and MacKenzie.[11] This performativity can refer to materialities which can be invoked, but which can also be dispensed with, in a large proportion of individual transactions (at least in the South African case) and systemically in the derivatives market as a whole. Value is instead discursively mobilised within a virtual framing of 'care' and the performativity of 'green'. But this virtuality is not without material effect – in fact far from it – in that it works to enable certain power holders to gain strategic access and control of natural resource assets and energy systems. But, most worryingly, this control and access is not necessarily reliant on any commensurate substantive switch to 'green' production or material practice. Instead, financial 'services' are enacted, which are the source of escalating rents from arbitrage,[12] for the benefit of owners of green funds, derivatives and increasingly bank-generated indexes – *in the absence* of substantive reductions in pollution levels.

Creating the Green Economy

Carruthers and Stinchcombe argue that, for liquidity to emerge, market participants need to establish what is standard or homogenous about the commodities they exchange. How this happens is a 'sociologically interesting process concern[ing] how heterogenous claims on income streams associated with different sorts of assets get turned into homogenous commodities that buyers and sellers can understand'.[13] Beckert, from within economic sociology, termed this 'the value problem', the 'processes of classification and commensuration with which actors assign value to goods'.[14] While many goods are sufficiently standardised to trade on public 'canonical mechanism' markets with regular trades and known and available pricing, goods such as asset-back security (ABS) collateralised debt obligations (CDOs) were traded 'over the counter', thus having similarities to early carbon trades and with goods that are less than standard.[15] Green bonds were also often bought by institutional investors from multilateral banks before the development of the indexes from mid-2014 to date, making the Green Economy, like the ABS CDO market, one in which analysis is required of the 'social processes behind the constitution of value'.[16] MacKenzie summarises areas in which canonical mechanisms do not easily apply. These include cases of contested prices; goods with incommensurable forms of evaluation; goods with high aesthetic judgements or when 'the quality of a product is inferred from the status of its producer'.[17] In other words, the Green Economy, composed of carbon trading initiatives, climate finance and other socio-technical arrangements of 'green' interventions, has many nodes where the expert rating or 'Gold Standard' of the issuer or project host is core to, or even determinant of, its perceived value. These are cases where economic sociology has a prescient role to play in understanding current experiments in market making, and especially the social processes of defining and agreeing 'value'.

Callon, and Callon and Muniesa describe the processes of framing, standardisation, and disentanglement and then re-entanglement that commodities undergo in order to enable this homogenisation.[18] Callon widens agency to include technologies, framings and evaluative processes within a socio-technical arrangement or 'agencement' within which the various projects of economists in the wild, traders, firms, marketing agencies, and so forth all compete and jostle for power over the way the eventual commodity emerges and is priced.[19] Miller critiqued this analysis by arguing that Callon was unnecessarily privileging the power of the abstract market of economic theory over evidence of the continued centrality of a 'highly entangled' world of exchange, citing his own theory of virtualism to describe 'the increasing ability of economists and other agents of abstract models such as audit and consultancy to transform the world into close approximations of their theory and models'.[20] His argument with Callon seems to be that he does not see this as inexorable or inevitable, whereas he depicts Callon as assuming it is (though maybe not correctly so, as evidenced in Callon's reply[21]). Essentially, according to Miller, 'for Callon, what lies within the frame is the market system, indeed it is the frame that helps define and preserve it as a market system. In my case – by direct contrast – what lies inside the frame is a ritual system, which is supported and defined by its opposition to that which lies outside the frame, which is the market system'.[22] The 'confusion' is that this ritual and ideological system is called 'the market' in Callon,[23] while all actual markets are instead understood as historical and ethnographic entanglements in Miller.[24] However, since Miller does not then further theorise what he means by 'inside' and 'outside', the distinction and difference between the two authors seems somewhat pedantic.

For our purposes both authors point to the power of evaluative practice in making markets: for Callon this is termed 'neoclassical anthropology' or 'anthropologics';[25] Miller sees neoclassical economics itself as essentially a form of anthropology and an ideological model associated with neoliberalism.[26] However, both see the evaluative practice as serving to make the world increasingly in its own image. This process will be explored here using the concept of performativity in recognition that both neoclassical anthropologics – ways of being and doing which embed a free market order – and participants' patterns of entanglement contribute to how the Green Economy is being made (which allows us to leave aside the intricacies of the Callon/Miller disagreement while borrowing the key themes from both, respectively). Performativity is also closely related to MacKenzie's use of 'reactivity', invoked in his remarkable work on how evaluative practice in the securities and derivatives markets in the 2000s contributed to the financial crash of 2008.[27] Here, the ratings agencies of securities became market makers, as traders conformed to their conventions on measurements of default, recovery, risk and most importantly correlation (the likeness of different underlying assets to react similarly to economy-wide risks) in order to achieve ratings that would attract investors. The calculation of risk related to ratings conventions was undertaken instead of employing more accurate mathematical models of actual structural risk to assets (although many claimed that the latter option was operationally and/or mathematically impossible in any case).[28] The 'evaluation practises crystallized in ratings reduce a difficult problem of evaluation…[of complex debt instruments]…to a simple one, by establishing a rough equivalence among debt instruments of different kinds and with different particularities'.[29]

Significantly to our argument here, however, the ABS CDO example explained by MacKenzie shows how the constructors of the instruments were able to anticipate the reaction to them from the ratings agencies and then construct key parameters, such as the rating and spread, in order to achieve certain ratings and thus buyer demand for the product. In other words, the design was informed by how ratings agencies would evaluate them, 'in a clear manifestation of what Espeland and Sauder (2007) call "reactivity": the effects of evaluation or ranking on what is being evaluated and ranked'.[30] In the case of ABS CDOs this evaluative practice allowed the quality of the underlying asset, most often mortgages, to become less significant than it should have been mathematically. This was in a context where securitisation had switched from a means to increase lending (to customers to buy homes) to a means to make profit through arbitrage, or 'lending in order to securitise' (so that traders could make a profit).[31]

In the case of carbon trading the parallel is that the value of the underlying asset, the dirty industry 'cleaning up' or sequestrating part of 'Nature', is of little temporal interest after the initial rating or certification, or scientific confirmation of carbon to be 'saved', has been made. Once 'rated' by the carbon experts, the certification by the UNFCCC (sometimes contested) finishes the presentation and packaging in order for Certified Emission Reduction (CER)[32] to be issued and entered into the secondary market of trading. In other words, there is a distinct new element of abstraction in that the product of evaluation, disentangled from place to form a CERs certificate, is as much a rating of the instrument to be sold, the 'project' denominated in tonnes of carbon, as it is an evaluation of the quality of the underlying activity or asset. It is the work of the experts at project level, carbon verifiers and validators at the Designated National Authority (DNA, the country institution managing the CDM), and at the UNFCCC, who contribute expertise, provenance and scientific authority which generates much of a CER credit's value. Thus this work from economic sociology has much to inform us of how we analyse valuation in the carbon economy.

The evaluation practices and persons developed in the carbon economy are also enjoying profitable roles in the emerging green bond market, with carbon trading and green bonds both being advocated as techniques that can contribute to the Green Economy. The green bond market also privileges what appear to be quite *ad hoc* evaluation and verification procedures, with the subsequent rating, 'Gold Standard', or inclusion in an index, having immense consequences for the demand for the bond or ABS. For example, of five current indexes for green bonds in 2015, one uses an NGO, Climate Bonds Initiative (CBI), one Bloomberg and the others in-house 'specialists'. We return in the fourth section to developments of evaluation practices generated in the CDM economy as they are carried forward into the green bond economy, where many new issues are also attracting the AAA ratings of Standard and Poor's and Moody's once enjoyed by the mortgage-based ABS CDOs up to 2008.

The Clean Development Mechanism in South Africa 2007–12

To understand performativity and the role of the virtual in the Green Economy we will first look at a case study of the South African CDM to explore how far evaluation itself makes the value in this domain. The CDM is a mechanism defined under the Kyoto Protocol of the UNFCCC that provides for carbon emissions reduction projects in Non-Annex I Parties (developing countries), whose contribution is measured in CER credits that can be traded in emissions trading schemes (ETSs), the most important of which has been the European Union ETS, allowing Annex 1 parties (industrialised countries) the ability to buy credits in order to meet their emissions caps. The CDM is supervised by the CDM Executive Board of the Conference of the Parties (COP) of the UNFCCC, and in-country by the Designated National Authority, which in South Africa is housed in the Department of Energy. The DNA, in turn, works to meet South Africa's emission mitigation activities defined in a 2010 letter to the UNFCCC as 'taking nationally appropriate mitigation actions to enable a 34% deviation below the "Business as Usual" emissions growth trajectory by 2020 and a 42% deviation below the "Business as Usual" emissions growth trajectory by 2025'.[33]

Between 2001, the first year of the scheme, and September 2012 the CDM had issued one billion CERs, with (by 1 June 2013), 57% of all CERs issued for abatement of HFC-23 (38%) or N_2O (19%).[34] In South Africa there were 54 CDM projects registered up to the country's 'graduation' and exit, as a result of increased sovereign wealth, from the CDM in 2012.[35] Where classification by industrial sector is determined by the direct parent company of the project, the mining, oil and gas, heavy metals and chemicals industries were the second largest CDM-funded group (enjoying 25% of projects numerically). Renewable energy and energy creation were the largest funded sector, with 61% of the 54 projects. The percentage of the share by activity or sector is produced in Table 1, calculated from the actual tonnes of projected abatement of CO_2 in the database for the projects listed. This should be treated with caution, however, since it is based on projected data for approved projects, and there is little official data on actual completion.

Thus, by September 2014, there were no official data on 38 of the 54 funded CDM projects in South Africa, which were marked 'status unknown/uncertain'; these included all the methane capture from landfill projects. Only eight projects

Table 1. Approved project CO_2 abatement by activity, South Africa (%).

Activity	% of total predicted CO_2 abated (tonnes)*
N20 abatement	22.42
Methane capture from landfill	16.16
Gas capture from closed ferrochrome furnaces	7.73
Biomass energy generation	3.45
Wind	33.15
Solar	6.14

*This table adds to 89.05% of a total of 9,651,395 tonnes in 52 projects of highly variable size, and excludes some projects not assignable to these categories.
Source: Total of all approved projects in the CDM database September 2014. https://cdm.unfccc.int.

were registered as completed successfully, with support withdrawn from two others by the UK Foreign and Commonwealth Office (Tugela Mill Fuel Switching Project) and the Government of Canada (Mariannhill and La Mercy Landfill Methane Capture). After exiting the CDM it appears that the oversight requirement of the DNA was not pursued, given the lack of public data, although five projects are recorded as in progress.[36]

There has been much critical material written of the CDM both globally,[37] and more specifically on South Africa,[38] most of the latter questioning the 'additionality' qualities of CDMs, given major spending on the minerals–energy complex, where additionality means that the project receiving funding via the CDM results in reductions in greenhouse gas emissions that would not have otherwise been achieved. For example, funding of a Sasol gas pipeline to a new brick kiln was justified by Sasol as saving carbon emissions that would have been generated if they were to pursue their 'original' intention of opening a coalmine. However, there was weak proof that the coalmine was ever going to be built in the first place.[39] In fact, of the 54 registered projects, 14 had extremely weak additionality cases with the other 40 mostly only moderately proved. Narratives of avoiding a dirty build (such as opening a coalmine) were weakly evidenced. Further, the 'additionality' or reduction in emissions appeared in around 12 projects to be already mandated by South African environmental law, most particularly the requirements of the National Air Quality Act of 2004.[40] In particular, section 33 of this Act on the rehabilitation of mines mandates companies to ensure post-closure air quality. However, 7.7% of the total predicted reductions in the South African database were committed to capturing waste gases from closed furnaces. The Sasol project was approved (representing alone nearly 10% of all abatement by tonnes of carbon) for abating its practice of releasing nitrous oxide into the atmosphere.[41] Both these cases problematise the idea of 'additionality' in relation to already existing Clean Air mandates. Similarly, CDM funding was provided to the Tongaat Hulett Fuel Switching Project, to reduce unpleasant smells, when this too is covered by section 35 of the Air Quality Act 2004. Thus, the overall evidence, at least for the industrial cases, supports the possibility that CDM funds are enhancing the commercial profitability of firms, by supporting projects that may have, or should have, happened in any case, such that additionality is weakly proved. The ongoing lack of public data and company monitoring reports available at the DNA also suggests that the status of CDM approvals lacks transparent oversight, even to the point where it is not clear if such approvals were implemented after CERs were released.

There is also a preference in evidence for the funding of projects by large-scale companies and multinationals, including Mobile Telephone Networks (Pty) Ltd (MTN), Omnia, Sasol, PetroSA, South African Breweries, Mondi, Denham Capital, and the Beatrix Mine. In the category of N_2O abatement, for example, which represents 22.42% of all the volume of potentially reduced CO_2 in the database, just three companies feature – Sasol, Omnia (a fertiliser company) and Africa Explosives Ltd – representing 9.95%, 8.51% and 3.96%, respectively of all projected CO_2 emissions abatements in the South African national CDM approved projects database. In terms of MTN, a multinational telecommunications group, CDM funding was provided for a 2.126 MW tri-generation plant (for electricity, cooling and heating) at its Johannesburg Head Office, which would switch its use of the national electricity grid to in-house energy fuelled from the Egoli gas pipeline. Since 92% of the electricity in the South African national grid is produced in coal-fired power plants, it was argued that 8617 tonnes of carbon emissions would be prevented.[42] The calculative logic of additionality included a calculation of the reduced impacts from the initial coal mining in terms of scarce water resources, SO_2 emissions and the disposal of coal ash. But there are several weak grounds in this justification. First, MTN is a global company, operating in 21 countries, which posted post-tax profits of more than ZAR41 million (in South Africa) in 2013, and could have self-funded the project. Removing the company from the risk of load-shedding by the national grid (when the electricity supply is suspended), its own reduction in ongoing electricity costs to Eskom, and the subsequent sale of its CERs to EDF, a major European electricity provider (mostly using nuclear) were also income streams MTN has enjoyed as a consequence of the CDM subsidy. Meanwhile, the Egoli pipeline is served by Sasol's Mozambican offshore gas reserves, whose emissions effects were not included in the calculation. This means that, overall, MTN, Egoli Gas, EDF and Sasol received public subsidy, despite being major polluters with the means to fund these changes themselves.

The initial positive concentration in the renewables energy sector (nearly 40% of the total predicted emissions reduction for wind and solar combined) also needs to be partially qualified, since ultimate beneficial owners of these firms were often established mining and energy companies or offshore mining, energy or infrastructure funds. For example, BioTherm Energy and Methcap received funding as renewables companies, but both are owned by Denham Capital Management (capitalised at US$4.3 billion) and PetroSA. The former has a global portfolio of largely mining and fossil-fuel energy concerns, and is also building, despite environmental protests, two 300MW coal-fired stations on Luzon Island in the Philippines, and exploiting virgin oil and tar sands in North America. Denham has also been sued several times for fraud, asset-stripping and aggressive business practices by company directors of firms it invests in (Tanner Cos and Vulcan Power Co). In addition, the category 'energy generation' includes seven landfill waste methane capture projects (representing 16% of the total database's predicted abatement), when evidence from Bisasar Road in Durban suggests that these schemes only lengthen the operating time of toxic dumps, while erroneously predicting the actual amount of methane that can be captured in practice.[43] There were also four projects designed to capture currently flared furnace off-gases, and a further four projects designed to recover closed furnace combustible waste gases, all designated as 'energy creation'.

This concentration of beneficiaries in the minerals energy complex is the material basis for international public subsidy, in the form of CDMs, to be captured by traditional fossil fuel and infrastructure funds offshore. Thus a nationally based power, energy and mining company has a 'parent' or equity fund owner in a secrecy jurisdiction. CERs and climate change mitigation funds are then co-mingled with fossil fuel investments in fund portfolios, thus indirectly cross-subsidising the worst offending dirty companies by increasing overall fund profits. For example, BHP Billiton has a stake in Coega IDZ Windfarm, while Denham Capital, as discussed above, owns part of BioTherm Energy and Methcap. In effect, the 'incentive' to invest away from fossil fuels is only marginally increased. This can either be viewed as public assistance for mitigation, as green fund managers would assert, or the production of token and marginal assets in order to perform a green categorisation for a much bigger economic entity. The replacement of old oil and mining equipment with a less emitting plant is considered green according to some analysts, while having a small proportion of green assets in a portfolio – termed 'non-pure play' – is deemed acceptable under a 'green' classification for a green bond (see below). In the sense that token assets or improvements can be used to reclassify a wider activity or firm as 'Green', the South African CDM case illustrates an evaluation practise common to, and further developing in, other Green Economy domains: in public-issued multilateral green bonds; in the Green Climate Fund (GCF): and within the private green bond market of 2015.

Green Economy valuation and green bonds

In the previous section it was concluded that receipt of public subsidy to reduce emissions under the CDM in South Africa was achieved with weak evidence of actual environmental improvements. This section concludes that this pattern of certification has also conditioned the growth of the Green Economy in respect to private funds, or 'green bonds', although techniques of proving additionality have been iteratively developed. While not a direct precursor, patterns of carbon evaluation have informed many of the evidence cases for what is a 'green bond', traded in the green bond market, although some green bonds are not related to the carbon economy directly. Issuance in the green bond market grew from $11 billion in 2013, to $34.2 billion on 20 November 2014, and is predicted to be at $300 billion by 2018.[44] The CBI estimates that in 2014 the 'total universe of bonds linked to key climate changes solutions' stood at $502.6 billion, compared to $346 billion in 2013,[45] with $7.2 billion in new green bonds issued in the first quarter of 2015. These figures appear large when compared to public OECD members' global expenditure on climate change of $9 billion in financial year 2013–14,[46] and, as of May 2015, to a Green Climate Fund which had generated $10.19 billion in pledges, of which $3.97 billion had been signed into contracts.[47] Meanwhile the total stock of global money was some $117 trillion in 2013, with some arguing that liquidity had reached much higher levels, with mineral fuels, including oil, coal, gas and refined products still making up 14.8% of all global trade.[48] In other words, green economy assets remain a small fraction of the global economy, and publically funded assets an even smaller component, encouraging many to look to the predominantly private market of green bonds for solutions.

As of July 2015 there were five main green bond indexes, where an index is used to track performance of green bonds that are included in the index: Solactive (the first index in March 2014), MSCI/Barclays (October 2014), two from Standard and Poor's (S&P) (one in July and another in September 2014), and Bank of America Merrill Lynch (BAML) (in October 2014).[49] Indexes improve liquidity, provide a benchmarking function and also theoretically allow for further derivatives trading, such as within synthetic indexes, and ABSs, ABS CDOs, and credit default swaps on indexes and synthetics.[50] The possibility of increased derivatives trading was enhanced by the launch of the world's first index-linked green bond fund, the State Street Global Green Bond Index Fund for institutional investors by State Street Global Advisors in April 2015, which tracks the Barclays/MSCI green bond index launched in 2014.[51] The indexes vary in size but are all growing rapidly, with most issues oversubscribed and the MSCI/Barclays viewed as the 'most green' as it includes post-issuance verification. Each index is different, having a 'unique selling point', summarised by Sean Kidney, CEO of NGO the CBI as 'It's all quality ice cream, but investors can still pick the flavour they want'.[52] Most significant are differing inclusion rules (for bonds to be listed on the index), which derive from the lack of standardisation in this relatively new market, something which in turn hinges on differing evaluations of 'what is green?' – there is 'no standard definition' nor 'resolution [of the problem] expected any time soon'.[53]

In the absence of market rules 'green' is currently determined by two main qualifications: either the proceeds of the bond are (supposed to) be ring-fenced for environmentally beneficial projects – called 'use of proceeds' bonds; and/or the issuers themselves badge them as 'green' with an accompanying narrative – called 'self-labelled' bonds.[54] Some indexes, like the carbon traders before them in the 'Gold Standard' brand, use external market verifiers. Solactive uses the Carbon Bond Initiative (an NGO) to decide what is 'green',[55] and includes self-labelled use of proceeds issued above $100 million in size, but excludes asset-backed notes, that is, securities (which meant that Toyota's $1.75 billion ABS to support leases for its hybrid cars was excluded), and is agnostic on project bonds.[56] The CBI provides a universe of approved issues from which the index's constituents are drawn, and Solactive then assesses them on the quality of the issuers' own explanation for the 'use of proceeds', with 65 included by October 2014, whose identities are not disclosed except to customers. Meanwhile Barclays and MSCI have written their own rules and methodology, and are still debating eligibility thresholds for large hydro-projects, annual reporting requirements and the 'use of proceeds' category.[57]

The S&P Dow Jones has two indexes, one for labelled use-of-proceeds of bonds (S&P Green Bond Index), and the other for project bonds (S&P's Green Project Bond Index). It assesses the 'use-of proceeds' narratives in both indexes to decide on inclusion. The latter included 21 bonds (as of November 2014), which were products of special purpose entities, asset-backed securities linked to green assets, or corporate bonds from 'pure-play' firms, where 'pure-play' means all the money is invested in the thing that is 'green' rather than just a part of it (which is termed 'non pure-play'). Thus in the latter there is 'no self-labelling requirement, but proceeds must be used entirely for green projects, and notes must be serviced by revenue generated by those projects',[58] generally over a 20-year period with an average size of just under $400 million in a single bond. Meanwhile, for consideration of inclusion in the BAML's Green Bond Index, a bond must first be tagged as green by Bloomberg (190 bonds were in 2014), then BAML uses the use-of-proceeds criterion, with reference to the Green Bond Principles (first launched in 2011, revised from January 2015), to select 51 bonds (as of November 2014), collectively valued at $31 billion and accounting for 77% of the overall capitalisation of the market.[59]

In practice, many assets designated as 'green' would be of dubious quality to environmental scientists. For example, in the HSBC Global Climate change

Index (CCI), there are 'pure play' companies, which are 'those deriving more than half their revenues from climate-related activities' and 'non-pure plays' which have presumably less revenue invested in this way, although down to what benchmark remains opaque. Also, a history of controversial 'green bond' investments is already growing, with early examples notably including GDF Suez's 'green' bond that financed the controversial Jirau Dam in Brazil which, with its neighbour Santo Antônio Dam, both on the Madeira River, the largest tributary of the Amazon, had 'serious and perhaps irreversible impacts on freshwater ecology and nearby communities'. It also threatens the survival of migratory fish, some indigenous communities in the area were previously un-contacted by the outside world, there are toxic effects on the river and flooding has been exacerbated, leading to thousands of displaced persons.[60]

When the quality, existence or materiality of assets are considered as epistemologically separable, it becomes clear that answering the question of what is, and what is not a green economy asset becomes impossible, not least because this judgment is a product of differing evaluation processes. From the above, particularly the South African case study, it can be established that some 'virtuality' exists, in that some projects or industrial improvements simply do not get implemented or do not exist, which is supported by evidence of fraud in the carbon economy more generally.[61] There is also a form of virtuality that can be isolated in reference to profits made in pure arbitrage opportunities based in the reinsurance and credit default swap markets of green bonds, supported by the fact that nothing has actually changed in the productive economy outside the world of derivatives. However, it is obvious that not all economic activities invested in by green stock are fictive. Without a research effort to 'drill-down' into the portfolio investments of bonds, which is beyond the scope of this paper, it cannot be established how far investments are 'real' or 'performed' when benchmarked against their real existence or scientific contribution to averting climate change.

Instead, the evidence above indicates that the real and virtual are co-produced through evaluation practice (even if we cannot establish to what exact degree they mix). This conditions how far investments cause changes in the material world, and how far the asset comes into existence as a consequence of a written or discursive framing and an associated life-world which remains virtual. The centrality of evaluative practice began within the carbon economy,[62] as the issuance of CERs introduced the idea that a derivative income stream (initially denominated in carbon) could attach to a green asset, such as a forest. A self-labelled or use-of-proceeds green bond develops this further, and could also promote a re-liquidation and cross-investment with assets from the carbon economy and within the commitments of the Green Climate Fund (GCF), suggesting that in future the green economy will become a complex arrangement across these investment nodes: CDMs, the GCF and green bonds. For example, a green bond does not have to own the underlying asset, but could issue an ABS against the Reducing Emissions from Deforestation and Forest Degradation (REDD+) project, which would give it a right to the income stream. But the 'asset' stays with the people who live there and have to in some sense pay for carbon sequestration (by changing their behaviour). This suggests an element of mortgaging, in that the asset now has to generate revenue to someone who has invested in it, or loaned an amount to it, whether or not this loan was taken on with choice or foresight or not by the proximate parties. A use-of-proceeds clause, evaluated by distant ratings agencies could easily become a cheap and unsatisfactory palliative, such as the provision of a small social programme, compared to the control of the forest that the bond would gain.

In mortgage ABS default, the asset would be 'recovered' to the bank, with the likely recovery ratios built in to the rating process. But would REDD+ forests and their residents be expected to give up their forest to the bank to help liqui-date a default position? Or would the asset framed as the 'project' need 'improv-ing' by the generating of more and better revenue streams for the green bond and credit default swap investor? The consequences are a probable deepening in external governance. Thus, while the late 2000s crash of the carbon trading mar-kets showed some traditional flaws in an engineered canonical exchange market (with a collapse of demand and oversupply), arguably another weakness did not fully emerge as a result of historical contingency (other weaknesses came to the fore first). Said weakness is that there is no strong institutional reason for the trade in carbon to reflect back on the quality of the underlying asset, except through weak and indirect routes of 'reputational risk' to the UNFCCC or inves-tors. But these were largely overcome by the reliance on and belief in certifica-tion systems. Arguably calculative and evaluation practices begun here, and more specifically, the experience of a greater, and seemingly sustainable, institutional distance between asset and income stream, prefigure the more abstract 'Green Economy' and 'green bond'. The quality of the asset (dirty factory or mine, for-est, dam) and its relationship to a derivative income stream in the carbon econ-omy (subsidised, certified CERs) becomes even less transparent and performed if a green bond or ABS is raised in reference to it. The owners of these have little incentive to check on the quality of the underlying asset if the certification sys-tem itself creates the asset in question. In short, it is not a forest anymore, but a REDD+ project; not a river, but an ecosystem service, and so forth. Although the 'proceeds' of the green bond may have to be returned to the 'project' invested in, this is only after management service fees, proceeds to investors and so forth have been extracted. The bond structure ensures high reward for those managing it, but little return or concern for the asset, environment or people in question.

In other words, evaluative practices have generated a situation in which the amount of value fixed in productive activities in the 'Green Economy' can remain very low, as evidenced by the figures at the beginning of this section, but liquid money in the 'Green Economy' can rise rapidly, reflecting the growing gap between these two categories of value – fixed in production, or circulating in exchange – in the global economy as a whole.[63] This liquidity generates other activities, employment and profits for verification and ratings agencies, fund managers and carbon traders, and bank employees, as well as fees for all the above and investors, before and sometimes instead of generating a proportionate increase in value actually invested in production and in averting climate change.

The political economy of value

Understanding the co-production of value within calculative entities (evaluation narratives, additionality calculations, and so forth) and in accumulation has precedents in generic value theory.[64] This work can help to analyse how far performativity, which is understood here as the role of the evaluative calculation in producing value purely on its own account, has become a systemic process or institution of the overall Green Economy assemblage. Assemblage here is taken to mean that agency has become extended in the institutional arrangement of the green economy to include the knowledge and power attributes of evaluation and technology. The Green Economy, beginning in the carbon economy, uses additionality narratives to entrain or entangle normativity in what is sold, making emotion central to the generation of price,[65] with the additional feature that the evaluation process creates a certification that co-produces value alongside the actual asset and economic activities around it. This allows value to circulate in derivatives at a scale increasing in respect to value embedded in production, just as predicted in general value theory.

Marx first theorised that the circulation of money as capital could be 'an end in itself' outside of production.[66] Polanyi then used 'fictitious capital' to describe 'any commodity whose social, cultural and/or ecological value exceeds the market value placed upon it within a capitalist system.[67] It has what Polanyi termed 'doubleness' or 'duality', inhabiting 'a world both within and beyond "the market"'.[68] Harvey developed the concept of 'fictitious capital', derived from 'credit money', and defined it as 'money that is thrown into circulation as capital without any material basis in commodities or productive activity'.[69] Some part of the green bond economy of fund holdings would be commensurate with these definitions, given the empirical evidence above.

The idea that circulation of value can form a system disconnected from units of production has also been developed by writers who focus on the economic anthropology of the work process.[70] For example, LiPuma and Lee argue that 'connectivity itself has become the significant sociostructuring value, leading to the emergence of circulation as a relatively autonomous realm, now endowed with its own social institutions, interpretative culture, and socially mediating forms'.[71] Meanwhile Büscher concurs with Graham that Marx's theory of value still forms the 'deep structure' of capital, but that 'Value production...has become more obviously "situated" in the valorized dialectics of "sacred" and powerful institutions, such as legislatures, universities, and transnational corporations...located today predominantly in "expert" ways of meaning and, more importantly, in their institutional contexts of production'.[72] These expert ways of meaning and institutional contexts of production seem to promise a post-production economy, with value embodied in 'reports, policy briefs, think tanks, brands, marketing, and so on',[73] such that 'what we call "values" are more or less ephemeral products of evaluation', which 'like all aspects of meaning...are socially produced and mediated'.[74] In this process evaluation techniques pursue

> what Bruno Latour calls theatres of persuasion, staging performances, amassing far-flung inscriptions and visualisations of all kinds in a single framework, and constructing technologies for comparing and manipulating them.[75]

Such efforts are a part of the current construction of the green economy, in the CDM and green bond market, as its participants promote verification, standardisation and rule-making in the creation of assets calibrated in environmental care and repair.

However, many contemporary Marxists, such as Harvey or Smith, retain the idea that 'the circulation of value requires also a physical circulation of material objects in which value is embodied or represented'.[76] In the case of the green economy, this would mean that attracting profit requires attachment or embedding in a material regime of accumulation: it is the overall assemblage which guarantees command over resources and which, as a holistic whole, expropriates profit from the commoner.[77] The material entrainment of people and nature is primary, such that the UNFCCC, the Gold Standard, or the CBI can assign accreditation and value only because local power-holders have already authored the project according to the particular power structure of the political economy in place, wherein local powerful actors are in support, project subjects are disciplined, and valuators are reacting to, as well as helping to form, social and ecological relationships. In this the evaluation technology is not the central determinant of value but co-produces it alongside these political economy relationships. As Lohmann reminds us, it is not the calculative technology *per se* that makes new exploitable zones for capital, since the objects (being made into assets) were not appropriated purely because they were subject to measurement, nor were they 'imaginary objects created by a reductionist "politics of knowledge" involving numbers and calculations...Rather, they helped constitute those relations'.[78] While Harvey and Smith insist on materiality as a Marxian law, Lohmann, MacKenzie, Miller, Callon and others attempt to explain how the material and discursive comingle and co-produce value and its attendant profit extraction.

For example, in our CDM case above, the underlying asset that is 'made' is a valuation denominated in CO_2 equivalence of an act of pollution of methane, sulphur, nitric oxide and so forth, which is to be abated in the future. It was not priced before but, via carbon calculation, can now be turned into CERs, which can be traded and transformed into money. It is a conversion from harm to care which simultaneously conjures an underlying asset into the carbon economy. However, the evidence from the CDM case and from the European Union Exchange Trading System (EUETS) more generally suggests that there is a limited demand for such products. This lack of investment in 'green' assets suggests that 'liquid nature', to borrow Büscher's term,[79] or more accurately in this case 'liquid care', attracts few buyers, making it unlikely that it is the prime focus of value creation. Instead, as Lohmann suggests, it is the relationship between the new green 'thing' and its surrounding political economy which conditions how far changes in ecology and well-being will occur. More specifically, when a Green Economy asset directly guarantees access to scarce, exploitable resources in a more conventional mode of accumulation, acting in effect as a 'gateway asset', it co-produces derivatives profits. For example, a REDD + generated carbon mitigation unit may struggle to find a buyer, since 'care' itself is a luxury commodity whose purchase lacks immediacy or compulsion, whereas ownership of the REDD + carbon mitigation unit, as part of a wider forestry assemblage,[80] could quite readily generate for its owner other derivative income streams, such as access to land and 'renewable' timber concessions where the REDD + scheme acts as offset. Ownership of a carbon sequestration project (added to the CDM structure in 2011) can generate overall control over timber supply and competitor suppliers. Similarly, the wind farms in the CDM portfolio in South Africa gave energy conglomerates access to public subsidy and control over their 'competitor' energy substitutes, and thus leverage on the energy regime as a whole. In this the Green Economy asset and income stream is but one of a cluster of resources within a political economy assemblage – be it forestry, agriculture, mining – but a determining one for overall control into the future as it embodies legitimacy for exploitative corporates, and/or has a gate-keeping role in relation to more lucrative assets.

The Green Economy and the virtual dimension

In Büscher's work on neoliberal conservation the poor and nature were depicted as '"underlying assets" for what has become the "real" source of value [...] namely images and symbols within the realms of branding, public relations and marketing'.[81] In his later work he argues for a 'liquid nature' circulating with even less relationship to any material fixed asset.[82] In contrast, Lohmann is much more adamant that the act of abstracting new nature is permanently messy and subject to continuing struggle as humans and non-humans, actants and things provide for resistance and subsequent changes in meaning and practise.[83] It is not possible to find a calibration by which we can judge abstraction to have taken place, and perhaps Büscher's characterisation is true of the most successful experiments, whereas Lohmann's is more applicable to describing the historicity of human–ecology relations and their vulnerability to new forms of resistance.[84] However, Igoe's insights on how abstraction is performed through spectacle,[85] do allow us to sidestep the epistemological exercise of attempting to calibrate an empirical scale from materiality to virtuality, which would prove unsatisfactory because it would inevitably force linearity on to complexity. Instead, we can ask a better question of how the material and virtual (non-existent and discursive) exist in relationship to each other. We have established above that the role of evaluative practices performs part of the translation of value between the two: a classification of 'greenness' can increase the value of a material asset, while the economic profitability of a material asset contributes in turn to its evaluation as deserving of the category of 'green'. The complexity of this co-production of value is found in a particular social anthropology.

Igoe, following Debord, uses the term spectacle as 'transform[ing] fragments of reality into a visually pervasive totality, produce[ing] "a separate pseudo world" (thesis 2), offered in exchange for the totality of actual activities and relationship, a world of "money for contemplation only" (thesis 49)'.[86] Broadening spectacle beyond image alone, it can be considered 'part of the wider mosaic that Michel Foucault...called techniques and technologies of government'.[87] In particular, Igoe isolates 'integrated spectacle' as of particular use to understanding the spectacle of conservation,[88] that is 'spectacle that has integrated itself into reality to the same extent that it was describing it, and that it was reconstructing it as it was describing it'.[89] This connects conceptually to the ideas of performativity in economics and to the notion that the 'value' of certain Green Economy assets and products is as much conjured by marketing, narrative and integrated spectacle as it is from a straightforward economic assessment of the underlying resources.

In order to view the Green Economy as a product of integrated spectacle, it is worth remembering that there are 'four important antecedents to spectacle: 1) forgetting, 2) abstraction, 3) reifications, and 4) proto-exchangeability'.[90] Forgetting how a thing is made, came into being, or its earlier 'use value' function, is part of it becoming exchangeable, which precedes its reification into something more than its previous self, such as a commodity, which then assumes a degree of universality and can be exchanged for others. Thus, in Igoe's terms, the conservation park is a universal representation of 'ideal universal nature'.[91] Similarly the Green Economy can be viewed as the universal branding of a set of activities that enjoy some proto-exchangeability reflected in the portfolios, and exchangeability of assets, from the carbon economy, climate financed projects and green bonds and green bonds funds. Iconic images such as wind farms, grasslands, blue globes held in hands, have come to signify the exchangeability of green economy assets in green bond trading activity; depicting 'the new end of an "averaged", "flexible" nature', whose appropriation and repair become comparable along the same scale of value.[92] For Lohmann this averaged, 'liquid' nature now makes 'hitherto uneconomic activities presented as ecological protection...worthy of capital investment'.[93] As we saw above, this would include the valuation of averaged past and present pollution, in order to acquire finance to mitigate it in the present and future. However, the calm, serene images of Green Economy trade fairs belie the despoilment on which some 'cleaning' or 'mitigation' is being rendered, generating an illustration of 'Spectacle's ability to project unity and consensus where none actually exists'.[94]

Conclusion

The key to which Green Economy assets are predominantly virtual, narratively performative or substantive and material, and whether they generate derivative income at scale, is to be found in the nature of the overall assemblage in which they are situated, or, in Lohmann's terms, the particular entrainment of commoner/frontier extractive social relations that is generative of capitalist surplus.[95] However, the geography and temporality of trading Green Economy assets is re-territorialised away from the space and place of the actual activities which are referent, however weakly, to the asset. In financial terms a private investment market centred on the 'green' is in the making, and its derivatives and securities infrastructure is being built. However, history shows us that fund managers have little incentive to check the quality of the underlying assets of investment funds. Taken back to place and space, there is evidence of non-performing, degraded and simply non-existing wind farms, offsets, carbon-capture technologies and tree planting efforts. The case study of South Africa is a case in point. Thus the Green Economy spectacle occludes the current human–nature assemblage, and the way that this territorialises pollution in concentrated areas of the global South, while re-territorialising the performance of care and repair, and trade in pollution, in the North. The materiality of a fossil fuels-based global political economy, and highly chemical-intensive global production system, is occluded by an evaluation system promising improvements and solutions, this latter performed through the Green Economy spectacle. In this assemblage profits are not necessarily derived from production-based economic activity, but from the lack of it, or the denial of it: thus accumulation here is spatial, temporal and co-dependent on the performance of the non-material, the spectacle and the ephemeral. It is suggested here that the social construction of value, and the process of performativity that this generates in societies and economies, requires further theorisation. Economic value can be made with little or no relation to material assets, unless the reason to do so is deeply embedded in actual resource politics where holders of money are still expropriating from the commoner. How 'making up' value, and 'taking up' value from underlying assets are related remains a conundrum in value theory.

Funding

This work is based on research supported by the South African Research Chairs Initiative of the Department of Science and Technology and National Research Foundation of South Africa [grant number 71220]. Any opinion, funding and conclusion or recommendation expressed in this material is that of the author and the NRF does not accept any liability in this regard. The work also contributes to the research programme funded by the Leverhulme Trust [grant award RP2012-V-041].

Notes

1. Sullivan, "Banking Nature?"
2. Zadek, "Greening Financial Reform"; Bracking, "The Anti-politics of Climate Finance"; IPCC, *Climate Change 2014*; and Hallegatte et al., "Future Flood Losses."
3. Pacala and Socolow, "Stabilisation Wedges"; O'Neill, "The Financialisation of Infrastructure"; and Castree and Christophers, "Banking Spatially on the Future."
4. Castree and Christophers, "Banking Spatially on the Future."
5. IPCC, *Climate Change 2014*.
6. While the CDM, carbon trading, climate finance and green bonds are not commensurate things, they are here seen as distinct domains of economic activity within the loose umbrella of the 'Green Economy'. They have in common that they designate 'greenness' to economic activities and are experiments in financialising nature or concern for nature.
7. Debord, *The Society of the Spectacle*; and Igoe, "Nature on the Move II."
8. Bracking, "How do Investors Value?"; Bracking, "Secrecy Jurisdictions"; and Bracking, "The Anti-politics of Climate Finance."
9. For example, Nel, *Assembling Value in Carbon Forestry*.
10. Harvey, *The Limits to Capital*.
11. Callon, *The Laws of the Markets*; Callon, "Why Virtualism paves the Way"; Miller, *Capitalism*; Miller, "A Theory of Virtualism"; Miller, "Turning Callon"; MacKenzie, "Making Things the Same"; and MacKenzie, "The Credit Crisis."
12. Arbitrage refers to making a profit on trading financial derivatives, typically securities, which have a lesser outflow than the income on the pool of debt or bonds on which the security is raised, even though both refer to the same underlying assets.
13. Carruthers and Stinchcombe, "The Social Structure of Liquidity," 354.
14. Beckert "The Social Order of Markets," 253–254.
15. A 'canonical mechanism' market is where the commodities traded are standardised with continuous auctions and widely known prices. An ABS CDO is a packaged derivative in which an 'asset-backed security' (ABS) – itself a set of claims on the cash flows from a pooled set of underlying assets such as

mortgages – becomes part of the portfolio of a 'collateralised debt obligation' (CDO) – which similarly claims incomes from the assets of corporate debt. See MacKenzie "The Credit Crisis"; and Carruthers and Stinchcombe, "The Social Structure of Liquidity."

16. Beckert, cited in MacKenzie, "The Credit Crisis," 1780.
17. MacKenzie, "The Credit Crisis," 1781.
18. Callon, *The Laws of the Markets*; Callon, "What does it Mean?"; Callon and Muniesa, "Economic Markets"; and Callon, "Why Virtualism paves the Way."
19. Callon, "Why Virtualism paves the Way."
20. Miller, "Turning Callon," 218.
21. Callon, "Why Virtualism paves the Way."
22. Miller, "Turning Callon," 223.
23. Ibid., 224.
24. Ibid., 280.
25. Callon, "Why Virtualism paves the Way," 10.
26. Miller, "Turning Callon," 219, 224.
27. MacKenzie, "The Credit Crisis."
28. Ibid.
29. Ibid., 1784.
30. Ibid., 1786.
31. Cited from interview data in ibid., 1823.
32. A single CER credit is equivalent to 1 tonne of carbon dioxide which can be counted towards meeting Kyoto targets. Collectively known as CERs they provide a standardized emissions offset instrument.
33. Government of the Republic of South Africa, "State of Nation Address Debate."
34. CDM, "CDM Projects grouped in Types."
35. UNFCCC CDM Project Database, 2014, http://unfccc.int. See also Government of the Republic of South Africa, "South African CDM Projects."
36. Bradfield, "An Analysis," 62.
37. MacKenzie, "Making Things the Same"; Bumpus and Liverman, "Accumulation by Decarbonization"; Newell and Bumpus, "The Global Political Ecology"; Liverman, "Conventions of Climate Change"; Castree, "Crisis, Continuity and Change"; Bumpus, "The Matter of Carbon"; and Bailey et al., "Ecological Modernisation."
38. Bond et al., "The CDM in Africa"; Bradfield, "An Analysis"; "Eskom told to charge Companies"; and Carbon Trade Watch, "Trading the Absurd."
39. Bond and Erion, "South African Carbon Trading."
40. Bradfield, "An Analysis." For example, the N_2O reduction project of African Explosives Ltd.
41. Sasol project documents, cited on the UNFCCC CDM Project database.
42. CDM, "CDM Clean Development Mechanism."
43. Bond and Sharife, "Africa's Biggest Landfill Site."
44. "A Review of the Green Market."
45. CBI 'The State of the Market in 2014' 2014 .
46. OECD, "Aid Activities."
47. GCF, "Status of Pledges." The GCF is a new specialist entity of the UNFCCC dedicated to attracting and spending climate finance.
48. Index Mundi, "World Economy Profile 2014."
49. Two further indexes exist: Nikko Asset Management (from 2010, but restricted to World Bank issues) and Calvert (October 2014, but extremely broad). See "The Rise of the Green Bond Index."
50. A credit default swap (CDS) transfers credit exposure between parties. The buyer of the swap makes payments to the seller in a similar way to insurance. The seller agrees to pay off a third-party debt if this party defaults, such as in an ABS. The buyer of a CDS might be speculating on the possibility that the third party will indeed default or be hedging an exposure position. A synthetic index tracks an established index and offers arbitrage profits if it exceeds returns on it.
51. "State Street launches."
52. Cited in "Bond Investors Ponder."
53. "A Review of the Green Market."
54. "Bond Investors Ponder."
55. Ibid.
56. Project bonds are dedicated to a classified green activity, such as solar power, in their entirety but are not seen to generate much profit.
57. "Barclays and MSCI launch Review."
58. "The Rise of the Green Bond Index."
59. Ibid.
60. "Bond Investors Ponder"; Environmental Finance "Is GDF's Green Bond Issue Really Green."
61. Interpol, "Guide to Carbon Trading Crime."
62. MacKenzie, "Making Things the Same."
63. Harvey, *The Limits to Capital*; and Büscher, "Nature on the Move."

64. Bracking et al., "Initial Research Design."
65. This was discussed in the debate between Miller and Callon on 'Sophie buys a car', where consumption involves the simultaneous disentanglement and re-entanglement of the object transferred and the bringing into play of all sorts of diverse cultural, social and contingent emotions and significations. Miller, "Turning Callon"; and Callon "Why Virtualism paves the Way." Thus emotion is not absent from prior theorisation of market exchange, but the argument here is that the Green Economy relies on this as the central core of the value to be exchanged, often in the absence of an associated 'use value' or material exchange.
66. Marx, cited in Büscher, "Nature on the Move."
67. Polanyi, *The Great Transformation*; and Castree, "Crisis, Continuity and Change," 192.
68. Polanyi, *The Great Transformation*.
69. Harvey, *The Limits to Capital*, 95.
70. Igoe, "Nature on the Move II"; LiPuma and Lee, *Financial Derivatives*; and Boltanski and Chiapello, *The New Spirit of Capitalism*.
71. LiPuma and Lee, *Financial Derivatives*, 19.
72. Graham, cited in Büscher, "Nature on the Move," 25.
73. Büscher, "Nature on the Move," 25.
74. Graham, *Hypercapitalism*, 4.
75. Ibid., 5.
76. Smith, cited in Büscher, "Nature on the Move," 24; Lohmann, "What is the 'Green'?," 8; and McNally, "Beyond the False Infinity," 12.
77. See Lohmann, "What is the 'Green'?"
78. Ibid., 3.
79. Büscher, "Nature on the Move."
80. See Nel, *Assembling Value in Carbon Forestry*.
81. Büscher, "Derivative Nature," 259.
82. Büscher, "Nature on the Move."
83. Lohmann, "What is the 'Green'?"
84. However, there is a potential difference in meaning for the concept of abstraction here: for Büscher it is a Marxian depiction of reification, while for Lohmann it refers to thought categories and cognitive universalisms framing new human–nature constellations.
85. Igoe, "Nature on the Move II."
86. Ibid., 38.
87. Ibid., 39.
88. Ibid., 39–40.
89. Ibid., 40.
90. Ibid.
91. Ibid.
92. Lohmann, "What is the 'Green'?," 9 Citing Helm, 'Taking Natural Capital Seriously'.
93. Lohmann, "What is the 'Green'?," 11
94. Igoe, "Nature on the Move II," 43.
95. See also Lohmann in 'Green Growth: Ideology, Political Economy and the Alternatives', forthcoming.

Bibliography

Bailey, Ian, Andy Gouldson, and Peter Newell. "Ecological Modernisation and the Governance of Carbon: A Critical Analysis." *Antipode* 43, no. 3 (2011): 682–703.
"Barclays and MSCI Launch Review of Green Bond Index Rules." Environmental Finance, July 21, 2015. https://www.environmental-finance.com/content/news/barclays-and-msci-launch-review-of-green-bond-index-rules.html.
Beckert, J. "The Social Order of Markets." *Theory and Society* 38 (2009): 245–269.
Boltanski, L., and E. Chiapello. *The New Spirit of Capitalism*. London: Verso, 2007.
"Bond Investors Ponder Fifty Shades of Green." Environmental Finance, August 12, 2014. https://www.environmental-finance.com/content/analysis/bond-investors-ponder-fifty-shades-of-green.html.
Bond, Patrick, and Graham Erion. "South African Carbon Trading: A Counterproductive Climate Change Strategy". In *Electric Capitalism: Recolonising Africa on the Power Grid*, edited by D. A. MacDonald, 338–358. London: Earthscan, 2009.
Bond, P., and K. Sharife. 2012. "Africa's Biggest Landfill Site: The Case of Bisasar Road." *Le Monde Diplomatique*, April 29. http://www.democraticunderground.com/101626606.
Bond, P., K. Sharife, and R. Castel-Branco. "The CDM in Africa cannot deliver the Money." Ejolt, 2012. http://www.ejolt.org/2012/12/the-cdm-in-africa-cannot-deliver-the-money-2/.
Bracking, S. "How do Investors value Environmental Harm/Care? Private Equity Funds, Development Finance Institutions and the Partial Financialization of Nature-based Industries." *Development and Change* 43, no. 1 (2012): 271–293.

Bracking, S. "Secrecy Jurisdictions and Economic Development in Africa: The Role of Sovereign Spaces of Exception in Producing Private Wealth and Public Poverty." *Economy and Society* 41, no. 4 (2012): 615–637.

Bracking, S. "The Anti-politics of Climate Finance: The Creation and Performativity of the Green Climate Fund." *Antipode* 47, no. 2 (2014): 281–302.

Bracking, S., D. Brockington, P. Bond, B. Büscher, J. J. Igoe, S. Sullivan, and P. Woodhouse. *Initial Research Design: Human, Non-human and Environmental Value Systems: An Impossible Frontier?* LCSV Working Paper Series 1. 2014. http://thestudyofvalue.org/wp-content/uploads/2013/11/WP1-Initial-Research-Design-final.pdf.

Bradfield, J. "An Analysis of Approved Clean Development Mechanism (CDM) Projects in South Africa." University of KwaZulu-Natal, SARCHi Applied Poverty Reduction Assessment, Technical Paper Number 3, mimeo, 2015.

Bumpus, Adam G. "The Matter of Carbon: Understanding the Materiality of tCO2e [sic] in Carbon Offsets." *Antipode* 43, no. 3 (2011): 612–638.

Bumpus, A. G., and D. M. Liverman. "Accumulation by Decarbonization and the Governance of Carbon Offsets." *Economic Geography* 84, no. 2 (2008): 127–155.

Büscher, B. "Derivative Nature: Interrogating the Value of Conservation in 'Boundless Southern Africa'." *Third World Quarterly* 31, no. 2 (2010): 259–276.

Büscher, B. "Nature on the Move: The Value and Circulation of Liquid Nature and the Emergence of Fictitious Conservation." *New Proposals: Journal of Marxism and Interdisciplinary Inquiry* 6, nos. 1–2 (2013): 20–36.

CDM. "CDM Clean Development Mechanism Project Design Document Form." (CDM-SSC-PDD) Version 03. December 22, 2006.

CDM. "CDM Projects grouped in Types." *CDM Investment Guide.* UNEP Risoe CDM/JI Pipeline Analysis and Database. June 1, 2013. http://www.cdmpipeline.org/cdm-projects-type.htm

Callon, M. *The Laws of the Markets.* Oxford: Blackwell, 1998.

Callon, M. "Why Virtualism paves the Way to Political Impotence: Callon Replies to Miller." *Economic Sociology European Electronic Newsletter* 6, no. 2 (2005).

Callon, M. "What Does it Mean to Say that Economics is Performative?" In *Do Economists Make Markets? On the Performativity of Economics,* edited by D. MacKenzie, F. Muniesa and L. Sui, 311–357. Princeton, NJ: Princeton University Press, 2007.

Callon, M., and F. Muniesa. "Economic Markets as Calculative Collective Devices." *Organization Studies* 26, no. 8 (2005): 1229–1250.

Carruthers, B. G., and A. L. Stinchcombe. "The Social Structure of Liquidity: Flexibility, Markets, and States." *Theory and Society* 28, no. 3 (1999): 353–382.

Castree, Noel. "Crisis, Continuity and Change: Neoliberalism, the Left and the Future of Capitalism." *Antipode* 41, no. S1 (2010): 185–213.

Castree, N., and B. Christophers. "Banking Spatially on the Future: Capital Switching, Infrastructure, and the Ecological Fix." *Annals of the Association of American Geographers* 105, no. 2 (2015): 1–9.

Climate Bonds Initiative (CBI). *Bonds and Climate Change the State of the Market in 2014.* London: CBI and HSBC, 2014. http://www.climatebonds.net/files/post/files/cb-hsbc-15july2014-a3-final.pdf

Debord, G. *The Society of the Spectacle.* New York: Zone Books, 1995.

"Eskom told to charge Companies Fair Rates." *Business Report,* February 23, 2010.

GCF. "Status of Pledges for GCF's Initial Resource Mobilization (IRM)." Accessed April 30, 2015. http://news.gcfund.org/wp-content/uploads/2015/04/release_GCF_2015_contributions_status_30_april_2015.pdf

Government of the Republic of South Africa. "South African CDM Projects Portfolio." 2015. http://www.energy.gov.za/files/esources/kyoto/2015/South-African-CDM-Projects-Portfolio-up-to-19March2015.pdf.

Government of the Republic of South Africa. "State of the Nation Address Debate by Minister of Water and Environmental Affairs, B. Sonjica." February 15, 2010. http://www.gov.za/state-nation-address-debate-minister-water-and-environmental-affairs-b-sonjica

Graham, P. *Hypercapitalism: New Media, Language, and Social Perceptions of Value.* New York: Peter Lang, 2007.

Hallegatte, S., C. Green, R. Nicholls, and J. Corfee-Morlot. "Future Flood Losses in Major Coastal Cities." *Nature Climate Change* 3 (2013): 802–806.

Harvey, D. *The Limits to Capital.* Chicago, IL: University of Chicago Press, 1982.

"Is GDF's Green Bond Issue Really Green?" Environmental Finance, 2014. https://www.environmental-finance.com/content/analysis/is-gdf%E2%80%99s-green-bond-issue-really-green.html

Helm, D. "Taking Natural Capital Seriously." *Oxford Review of Economic Policy* 30, no. 1 (2014): 109–125.

Igoe, J. "Nature on the Move II: Contemplation becomes Speculation." *New Proposals: Journal of Marxism and Interdisciplinary Inquiry* 6, nos. 1–2 (2013): 37–49.

Index Mundi. "World Economy Profile 2014." 2014. http://www.indexmundi.com/world/economy_profile.html.

Intergovernmental Panel on Climate Change (IPCC). *Climate Change.* Synthesis Report, 2014. Nairobi: UNEP, 2014.

International Criminal Police Organisation (Interpol). "Guide to Carbon Trading Crime." Environmental Crimes Programme, June 2013. http://www.interpol.int/content/download/20122/181158/version/3/file/Guide%20to%20Carbon%20Trading%20Crime.pdf

LiPuma, E., and B. Lee. *Financial Derivatives and the Globalization of Risk.* Durham, NC: Duke University Press, 2004.

Liverman, Diana M. "Conventions of Climate Change: Constructions of Danger and the Dispossession of the Atmosphere." *Journal of Historical Geography* 35, no. 2 (2009): 279–296.

Lohmann, L. "What is the 'Green' in 'Green Growth'?", 2015. http://www.thecornerhouse.org.uk/sites/thecornerhouse.org.uk/files/GREEN%20GROWTH%20web%20version%204.pdf

Lohmann, L., "What is the 'Green' in 'Green Growth'?." In *Green Growth: Ideology, Political Economy and the Alternatives*, edited by G. Dale, Manu V. Mathai and Jose A. Puppim de Oliveira. London: Zed Books, forthcoming.

MacKenzie, D. "Making Things the Same: Gases, Emission Rights and the Politics of Carbon Markets." *Accounting, Organizations and Society* 34, nos. 3–4 (2009): 440–455.

MacKenzie, D. "The Credit Crisis as a Problem in the Sociology of Knowledge." *American Journal of Sociology* 116, no. 6 (2011): 1778–1841.

McNally, D. "Beyond the False Infinity of Capital: Dialectics and Self-mediation in Marx's Theory of Freedom." In *New Dialectics and Political Economy*, edited by R. Albritton and J. Simoulidis, 1–23. New York: Palgrave Macmillan, 2003.

Miller. D. *Capitalism: An Ethnographic Approach.* Oxford: Berg, 1997.

Miller, D. "A Theory of Virtualism." In *Virtualism: A New Political Economy*, edited by J. Carrier and D. Miller, 187–215. Oxford: Berg, 1998.

Miller, D. "Turning Callon the Right Way Up." *Economy and Society* 31, no. 2 (2002): 218–233.

Nel, A. *Assembling Value in Carbon Forestry.* LCSV Working Paper 10. 2015. http://thestudyofvalue.org/wp-content/uploads/2015/01/WP-10_Nel-2015-Assembling-value-in-carbon-forestry.pdf

Newell, P., and A. Bumpus. "The Global Political Ecology of the Clean Development Mechanism." *Global Environmental Politics* 12, no. 4 (2012): 49–67.

O'Neill, P. "The Financialisation of Infrastructure: The Role of Categorisation and Property Relations." *Cambridge Journal of Regions, Economy and Society* 6, no. 3 (2013): 441–457.

OECD. "Aid Activities targeting Environmental Objectives: Statistics." Accessed April 26, 2013 http://stats.oecd.org/Index.aspx?DataSetCode=RIOMARKERS.

Pacala, S., and R. Socolow. "Stabilisation Wedges: Solving the Climate Problem for the Next 50 Years with Current Technologies." *Science* 305 (2004): 968–972.

Polanyi, K. "*The Great Transformation." Boston, MA: Beacon* 2009 (1944).

"A Review of the Green Market in 2014." Environmental Finance, 2014 https://www.environmental-finance.com/content/analysis/a-review-of-the-green-market-in-2014.html.

"The Rise of the Green Bond Index." Environmental Finance, November, 6, 2014. www.environmental-finance.com

"State Street launches World's First Index-linked Green Bond Fund." Environmental Finance, May 8, 2015. www.environmental-finance.com

Sullivan, S. "Banking Nature? The Spectacular Financialisation of Environmental Conservation." *Antipode* 45, no. 1 (2013): 198–217.

"Trading the Absurd." Carbon Trade Watch, November 17, 2005. http://www.carbontradewatch.org/index.php?option=com_content&Itemid=36&id=177&task=view.

Zadek, S. "Greening Financial Reform." 2013 http://www.project-syndicate.org/commentary/integrating-the-green-growth-imperative-and-financial-market-reform-by-simon-zadek.

Index